Hooked on Math

Elementary Math

Steps 7 & 8

Workbook

This workbook and the included storybooks introduce beginning math skills.

Step 7: Place Value to 1,000
- Understanding place value to 1,000
- Working with place value sums

Step 8: Count and Compare Numbers
- Counting to 1,000
- Identifying counting patterns
- Comparing four-digit numbers

Hooked & Company

Contents

Lesson Overview

Welcome! This workbook provides supplemental activities that will solidify and expand the learning your child is enjoying with the Hooked on Math App. Follow the easy, step-by-step instructions to guide your child's exploration of numbers and operations. Every lesson helps your child build a strong foundation in elementary math skills while having fun!

1. Launch

Your child will be introduced to math concepts and skills through interactive lessons in the App on your mobile device or Windows or Mac computer. Download the App at:
hookedonmath.com/get-the-app

2. Discover

Next, your child will have fun exploring math concepts and practicing new skills in the workbook. This will reinforce the App lessons and provide a tangible application of the skills learned.

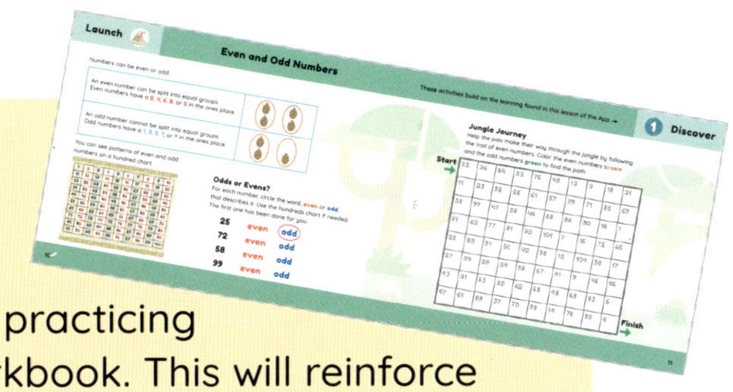

3. Play

In addition to the workbook activities, your child will enjoy two math-based storybooks that will strengthen their learning and add to the fun!

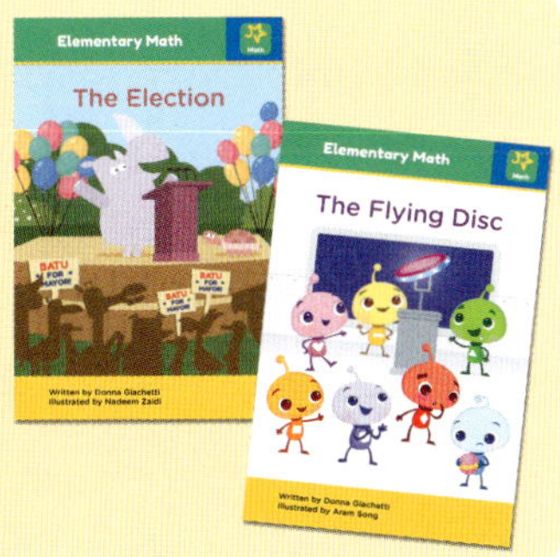

4. Celebrate

Celebrate your child's progress by placing a sticker in the workbook as learning milestones are reached. This also provides a helpful marker to remember where to begin next time.

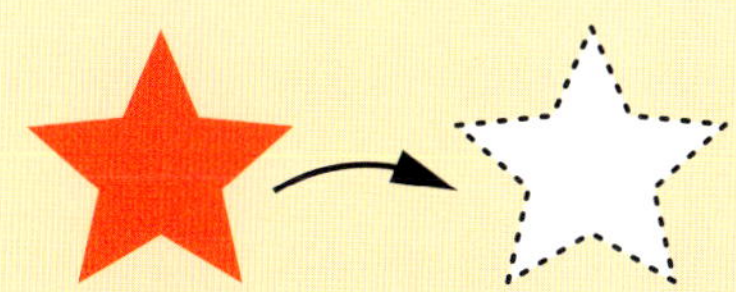

Special notes are included in this space to provide additional context related to what is being taught.

Counting Back on a Hundred Chart

See the Problem?

Each section of the hundred chart shows a subtraction sentence.

- Write the starting number. Remember that the larger number is always first.
- Count and write the number being taken away.
- Write the difference. The first one has been done for you.

51	52	53	54	55	56	57	58	59	60
61	62	63	64	65	66	67	68	69	70
71	72	73	74	75	76	77	78	79	80
81	82	83	84	85	86	87	88	89	90
91	92	93	94	95	96	97	98	99	100

64 - 8 = 56

51	52	53	54	55	56	57	58	59	60
61	62	63	64	65	66	67	68	69	70
71	72	73	74	75	76	77	78	79	80
81	82	83	84	85	86	87	88	89	90
91	92	93	94	95	96	97	98	99	100

___ - ___ = ___

1	2	3	4	5	6	7	8	9	10
11	12	13	14	15	16	17	18	19	20
21	22	23	24	25	26	27	28	29	30
31	32	33	34	35	36	37	38	39	40
41	42	43	44	45	46	47	48	49	50

___ - ___ = ___

51	52	53	54	55	56	57	58	59	60
61	62	63	64	65	66	67	68	69	70
71	72	73	74	75	76	77	78	79	80
81	82	83	84	85	86	87	88	89	90
91	92	93	94	95	96	97	98	99	100

___ - ___ = ___

1	2	3	4	5	6	7	8	9	10
11	12	13	14	15	16	17	18	19	20
21	22	23	24	25	26	27	28	29	30
31	32	33	34	35	36	37	38	39	40
41	42	43	44	45	46	47	48	49	50

___ - ___ = ___

51	52	53	54	55	56	57	58	59	60
61	62	63	64	65	66	67	68	69	70
71	72	73	74	75	76	77	78	79	80
81	82	83	84	85	86	87	88	89	90
91	92	93	94	95	96	97	98	99	100

___ - ___ = ___

Block by Block

Solve each subtraction problem by crossing out a tens rod and drawing it as 10 ones blocks. Then cross out the number of blocks being taken away. Write the subtraction sentence. The first one has been done for you.

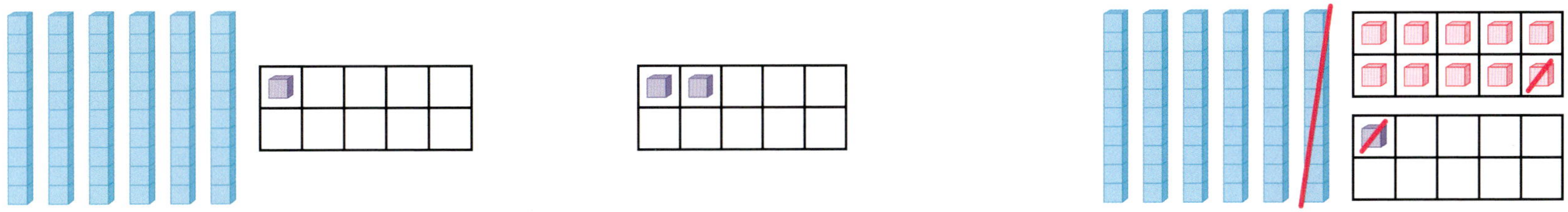

61 - 2 = 59

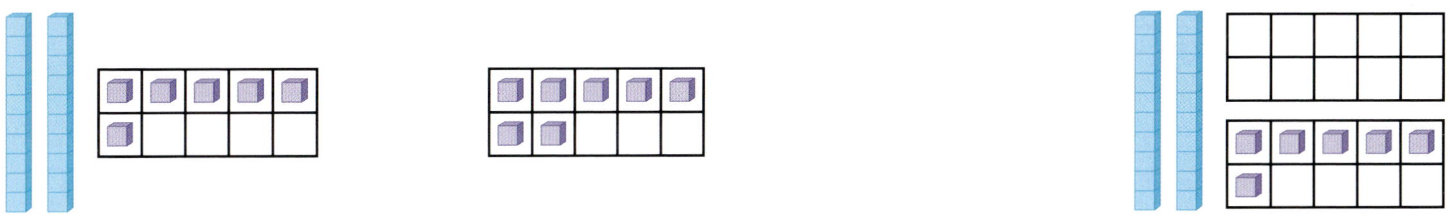

___ - ___ = ___

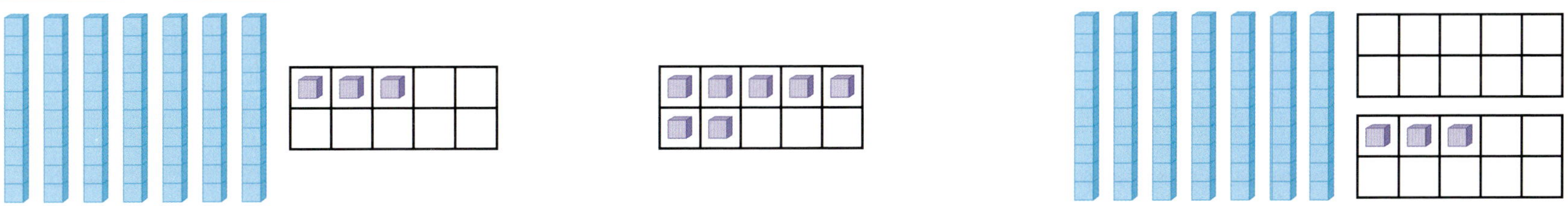

___ - ___ = ___

Vertical Subtraction with Regrouping

Go for a Spin

1. Spin Spinner 1 and write the number in the top box.
2. Spin Spinner 2 and write the number in the box below it.
3. Subtract and write the difference. An example has been done for you.
4. Place an **X** over that number if it appears on a BINGO card.
5. Keep spinning and subtracting. If the difference is a number that you have already marked with an **X**, spin again. Use scratch paper if needed.
6. The first card to get five numbers in a row, across, down, or diagonally, wins!

Spinner 1

Spinner 2

You can play alone or with a friend. If you like, play again with any extra boxes or scratch paper.

Tens	Ones
8 ~~9~~	14
- 3	7
5	7

Tens	Ones
-	

Tens	Ones
-	

Tens	Ones
-	

Tens	Ones
-	

Materials needed: sharpened pencil, paper clip

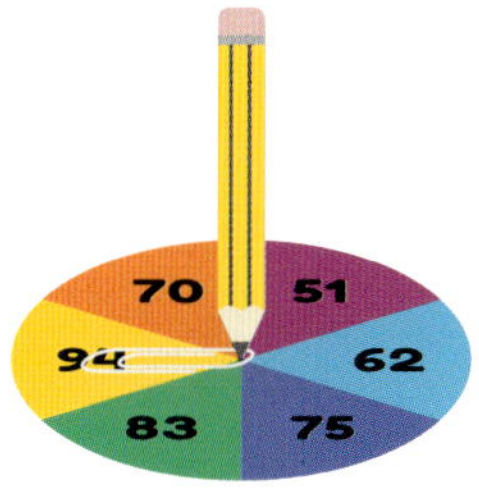

How to Make the Spinner: Place one end of the paper clip in the center of the circle. Put the pencil tip inside the paper clip on the center dot. Now flick the paper clip so it spins around the pencil.

B	I	N	G	O
2	25	36	44	52
13	26	38	45	~~57~~
14	28	Free	46	58
17	33	39	49	60
21	34	41	50	63

B	I	N	G	O
13	28	41	50	65
17	33	44	52	68
21	34	Free	~~57~~	71
25	36	46	58	76
26	39	49	63	82

Tens	Ones
−	

Tens	Ones
−	

Tens	Ones
−	

Tens	Ones
−	

Tens	Ones
−	

Celebrate finishing the review with a star sticker.

Place Value Sums to 1,000

A place value chart helps you to understand the value of each digit in a number.

Look at the value of each digit in this number: 6,347

The value of the number **6** is six thousands:	6,000
The value of the number **3** is three hundreds:	300
The value of the number **4** is four tens:	40
The value of the number **7** is seven ones:	+ 7
Added together, the sum is:	6,347

The mark between the 6 and 3 in the number 6,347 is called a *comma*. In numbers that are larger than three digits, a comma is inserted after every third digit, counting from right to left.

Thousands	Hundreds	Tens	Ones
6,	3	4	7

Don't forget to add the comma.

Try It!

Add the numbers. Write the sum in the place value chart.

Thousands	Hundreds	Tens	Ones

5,000 + 200 + 70 + 6

Thousands	Hundreds	Tens	Ones

9,000 + 600 + 80 + 9

Thousands	Hundreds	Tens	Ones

4,000 + 100 + 20 + 1

Digit Detection

Look at the sum in each addition sentence. Write the value of the missing number. The first one has been done for you.

8,000 + 600 + 20 + 8 = 8,628

______ + 500 + 40 + 3 = 4,543

2,000 + 900 + ____ + 6 = 2,956

1,000 + 300 + 70 + ____ = 1,371

______ + 400 + 30 + 2 = 5,432

3,000 + ______ + 10 + 8 = 3,718

9,000 + 200 + ____ + 4 = 9,294

Launch

Place Value Word Problems

Number Story

Place value problems can come in the form of a story, or word problem.

Read the math story. Circle the number you are working with and underline the value you are trying to find.
Write the number you are working with in a place value chart. Circle the value you are trying to find.
Write a sentence that shows the value of the number you are trying to find.

Chow blew up 62 balloons.
What is the value of 6 in the number?

Thousands	Hundreds	Tens	Ones
		6	2

The value is **60**.

Try It!

Kazoo baked 124 cupcakes for the party.
What is the value of 1 in the number?

Thousands	Hundreds	Tens	Ones
	1	2	4

The value is ________.

In Its Place

Read each word problem below. Then:

- Circle the number you are working with and underline the value you are trying to find.
- Write the number in the place value chart.
- Solve the problem and write the answer to the question. The first one has been done for you.

Read It	Write It	Answer it
There are 2,633 seats in the baseball stadium. What is the value of 2 in this number?	Thousands: 2, Hundreds: 6, Tens: 3, Ones: 3	The value is 2,000.
There are 379 students in the school. What is the value of 7 in this number?	Thousands: ___ Hundreds: ___ Tens: ___ Ones: ___	The value is ______.
Kazoo is tracking his steps. He has taken 9,264 steps today. What is the value of 2 in this number?	Thousands: ___ Hundreds: ___ Tens: ___ Ones: ___	The value is ______.
The volunteer group picked up 1,245 bottles and cans. What is the value of 5 in this number?	Thousands: ___ Hundreds: ___ Tens: ___ Ones: ___	The value is ______.

Launch

Place Value Blocks to 1,000

A number's place value can be shown with cubes, flats, rods, and blocks.

Look at the number 2,534.

The value of **2** can be shown by 2 thousands cubes.

The value of **5** can be shown by 5 hundreds flats.

The value of **3** can be shown by 3 tens rods.

The value of **4** can be shown by 4 ones blocks.

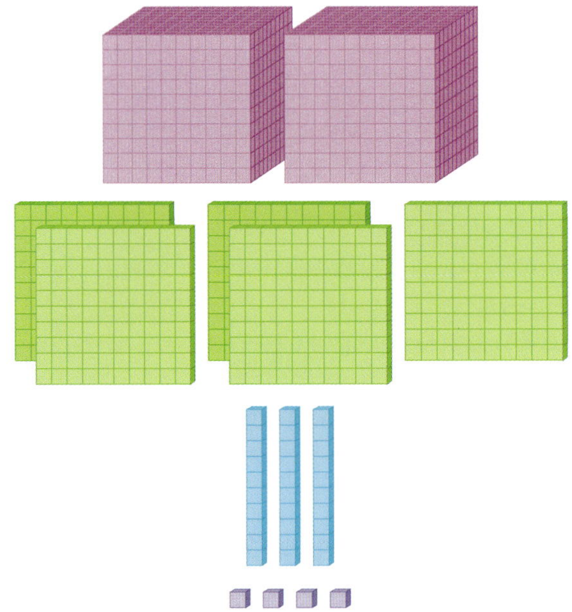

Try It!

Underline the digit in the number that matches the value in the picture. The first one has been done for you.

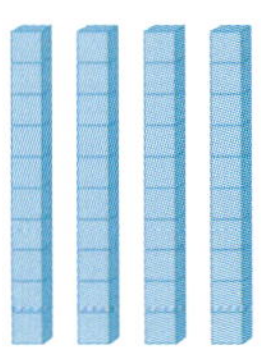

4,649

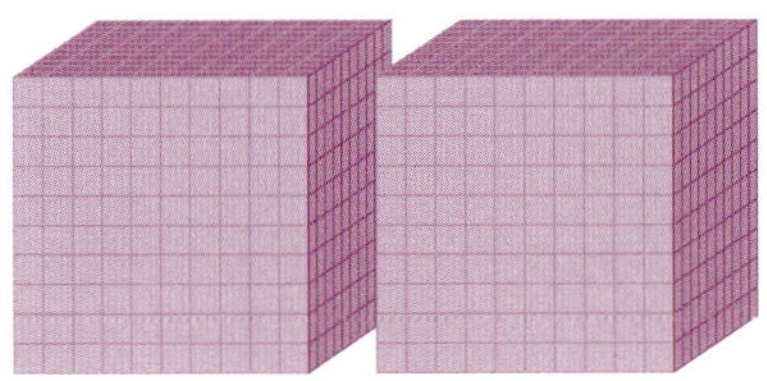

2,723

5,171

Don't Forget!

Cube = 1,000

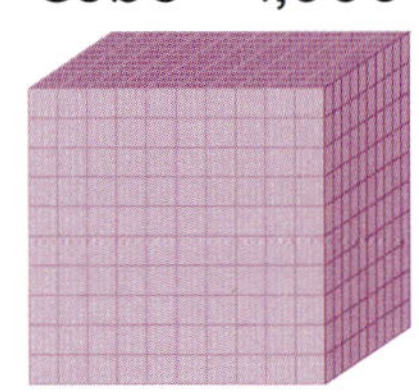

Flat = 100

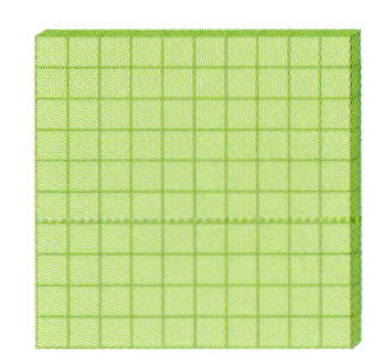

Rod = 10

Block = 1

How Many of Each?

Write how many thousands, hundreds, tens, and ones are in each number. Then look at the table to answer the questions below. The first one has been done for you.

- Which digit represents the most ones? Color it **yellow**.
- Which digit represents the fewest tens? Color it **orange**.
- Which digit represents a value equal to 400? Color it **green**.
- Which digit represents a value equal to 8,000? Color it **blue**.

7,256	7	2	5	6
9,832				
2,491				
8,574				

Materials needed: crayons or colored pencils in yellow, orange, green, and blue

Launch

Modeling Numbers to 1,000

An entire number can be shown, or *modeled*, using place value blocks.

How do you model the number 2,304?

There are **2** thousands cubes in 2,000.

There are **3** hundreds flats in 300.

There are no tens rods in **0**.

There are **4** ones blocks in 4.

Try It!

Draw a line to match the numbers to the correct place value block models.

3,056

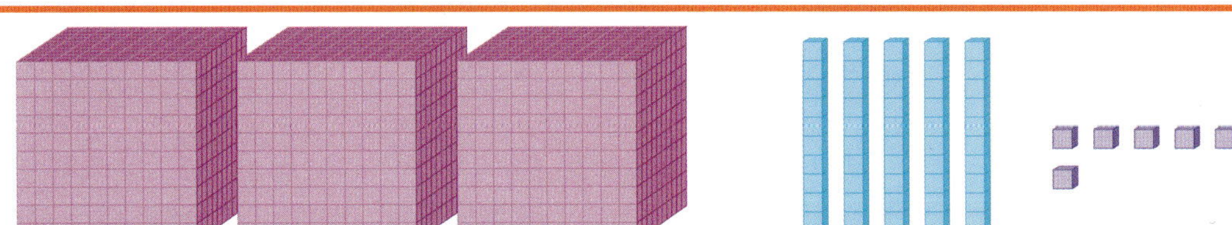

1,204

Put in Place

Look at the cubes, flats, rods, and blocks. Write how many each one represents on the line below it. Then add them together and write the sum. The first one has been done for you.

2,000 + 100 + 50 + 6 = 2,156

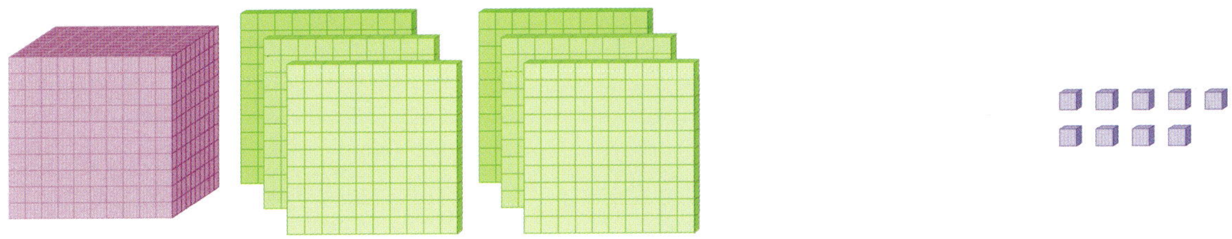

______ + ______ + ____ + ___ = ______

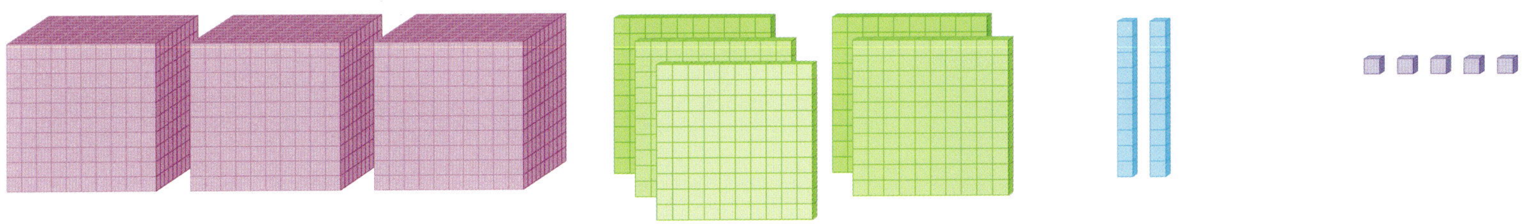

______ + ______ + ____ + ___ = ______

Representing the Same Number in Multiple Ways

The Same but Different

Usually, numbers are best represented with the fewest place value blocks possible.	135 = 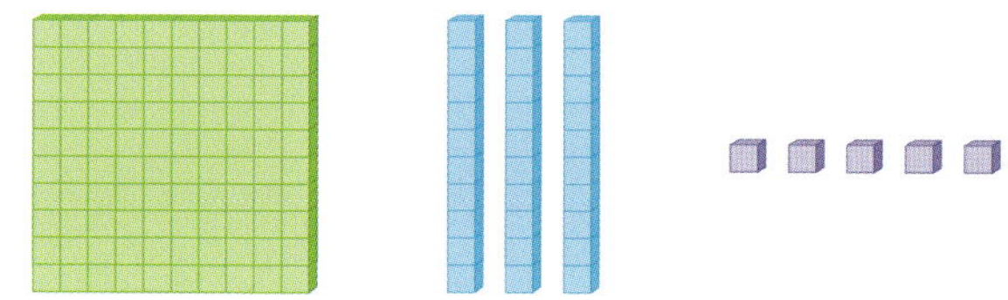
But sometimes, it is better to represent a number in a different way, to help with regrouping in operations such as addition and subtraction problems.	135 = 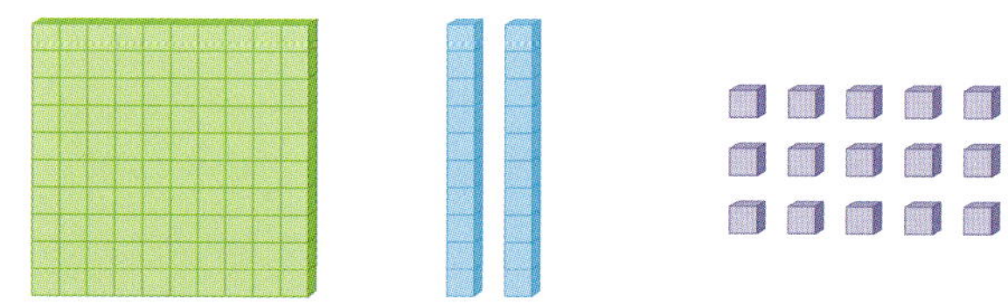

Try It!

Draw a line to match the place value blocks that represent the same number. Then match it to the numeral. The first one has been done for you.

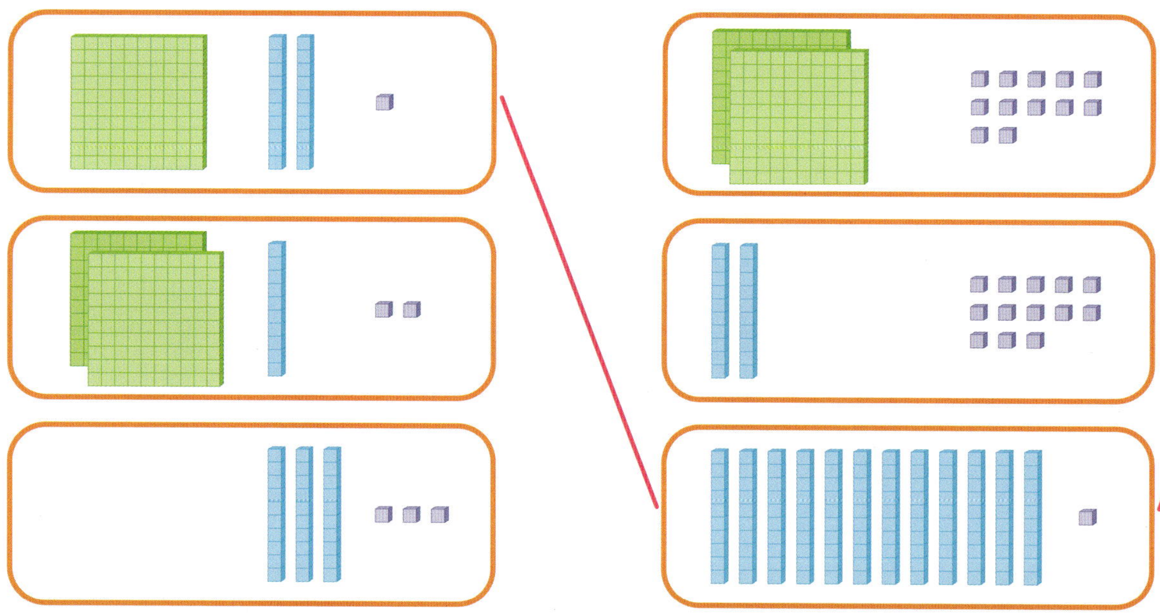

121

33

212

Which Doesn't Belong?

Look at the pictures next to each number. Circle a picture if it matches the number. Cross out a picture if it does not match. Hint: in some rows, both sets of pictures match the number.

1,276	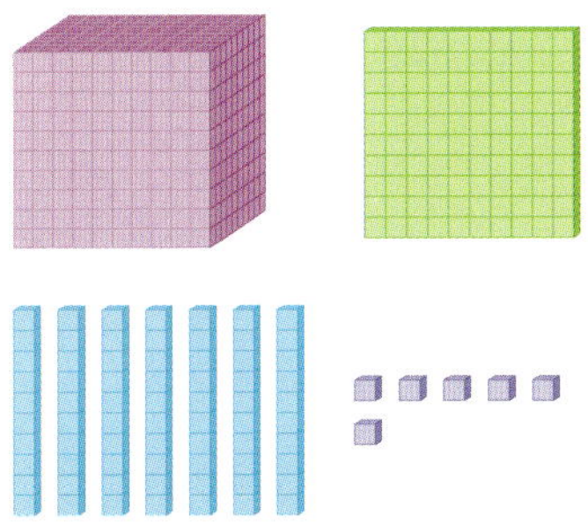	
431	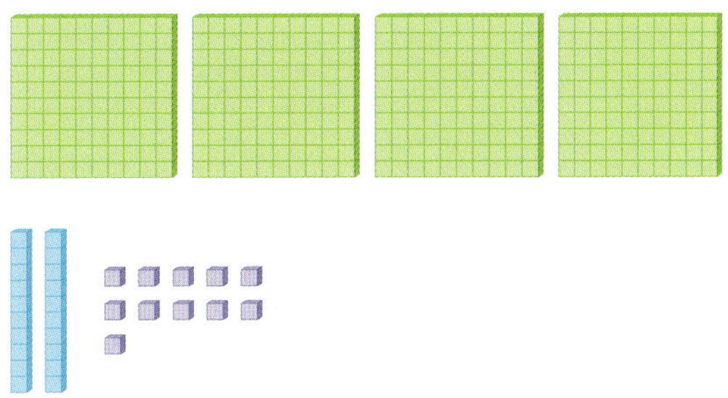	
916	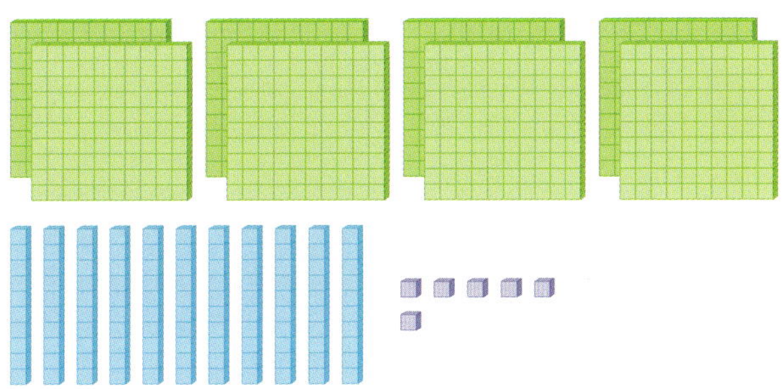	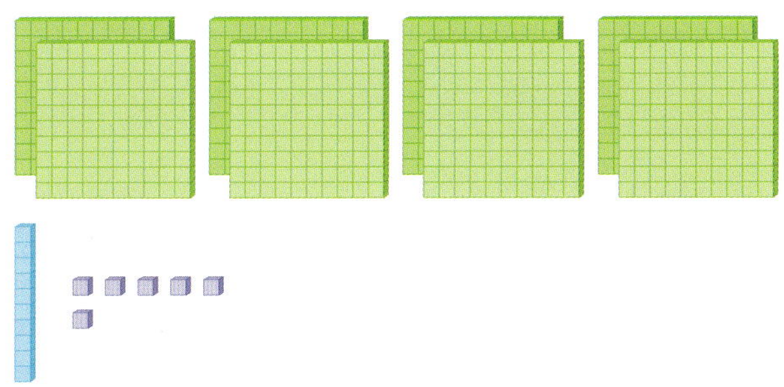

Play

A Valuable Color

Know the Number

Color the picture on the opposite page according to the Color Key.

Color Key

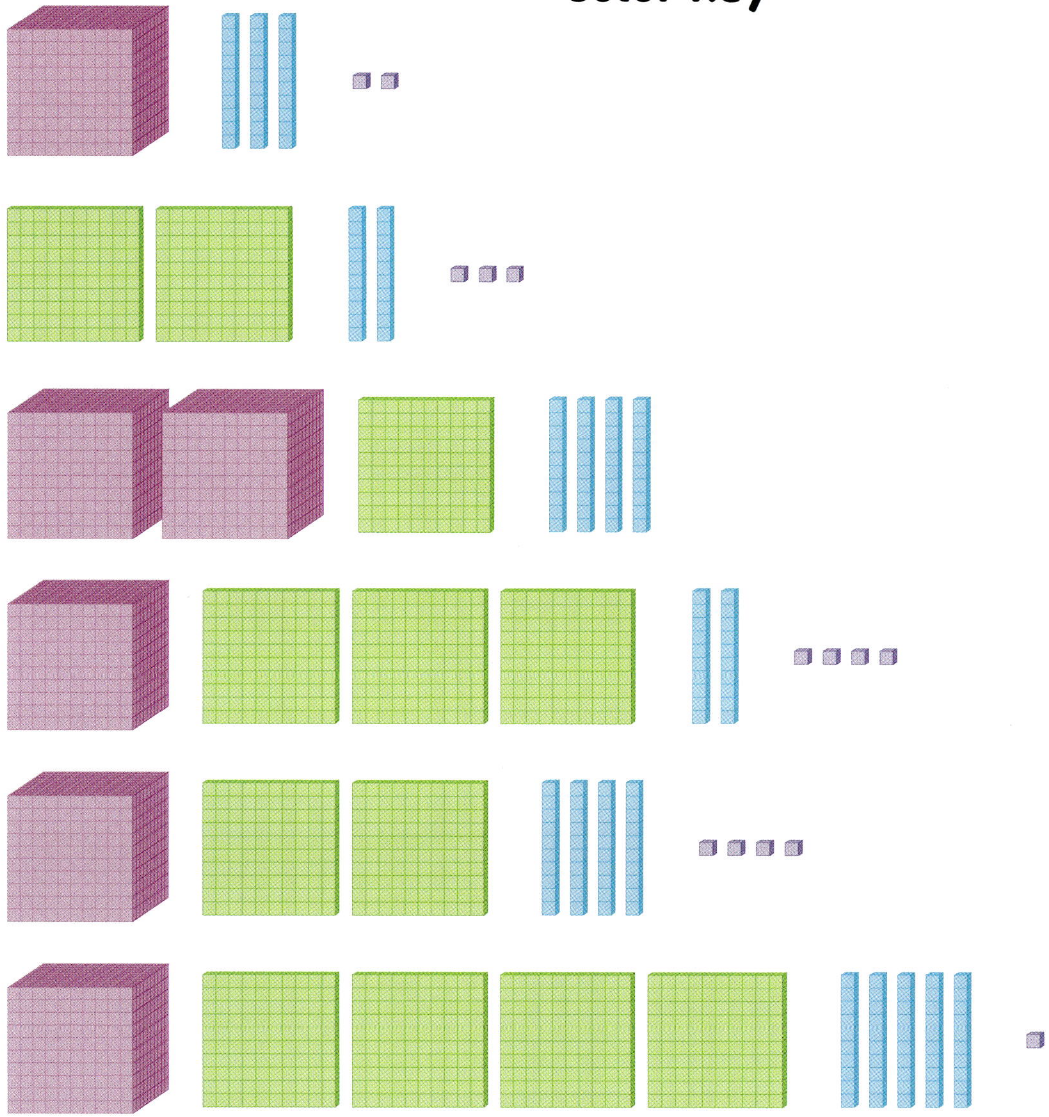

= red

= yellow

= purple

= green

= light blue

= dark blue

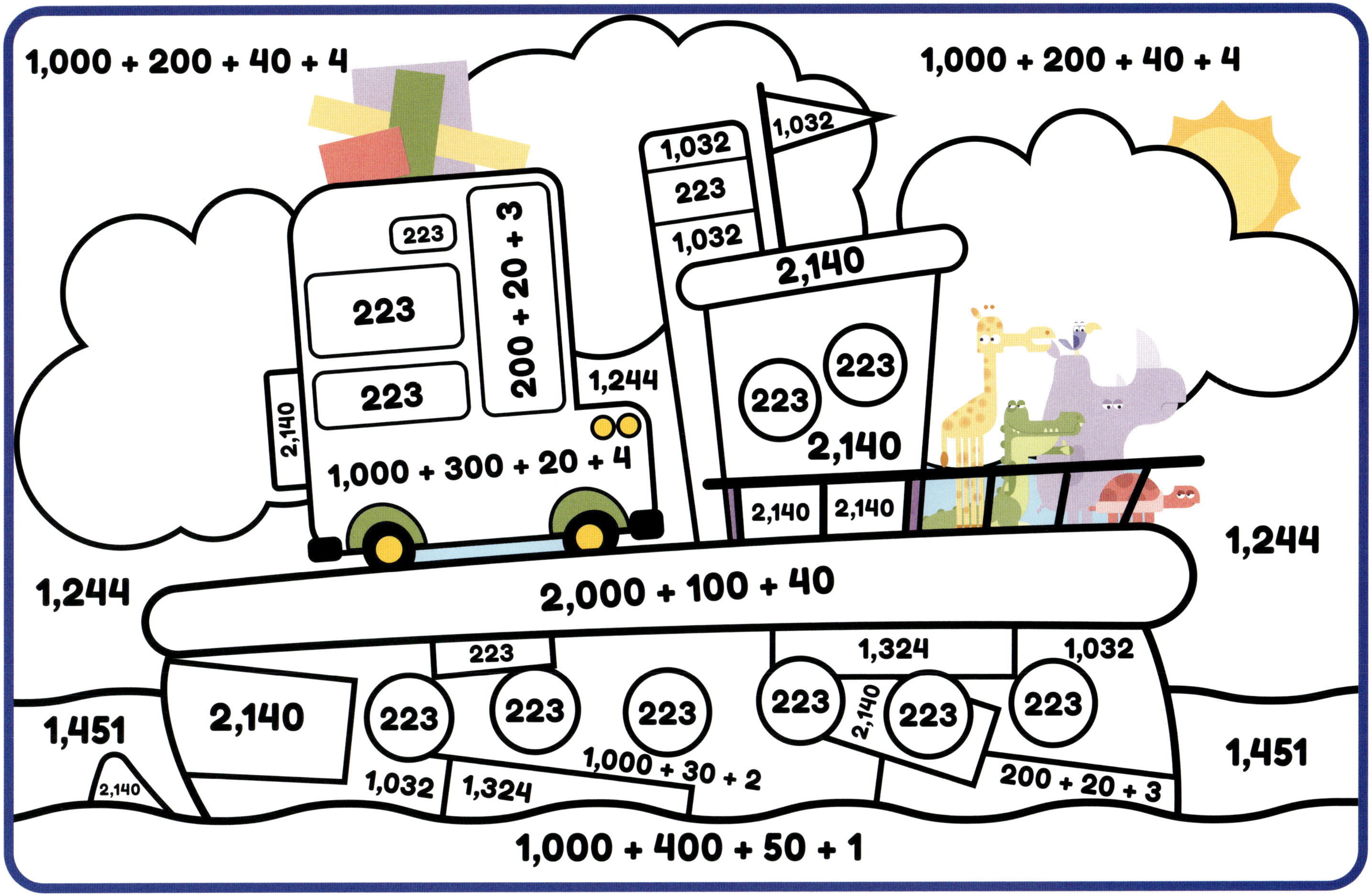

Materials needed: crayons or colored pencils in red, yellow, purple, green, light blue, and dark blue

The Election

Your child is ready to read *The Election*, a humorous story that features place value comparisons. Place value charts on some pages help your child compare the votes as they are tallied for each pal.

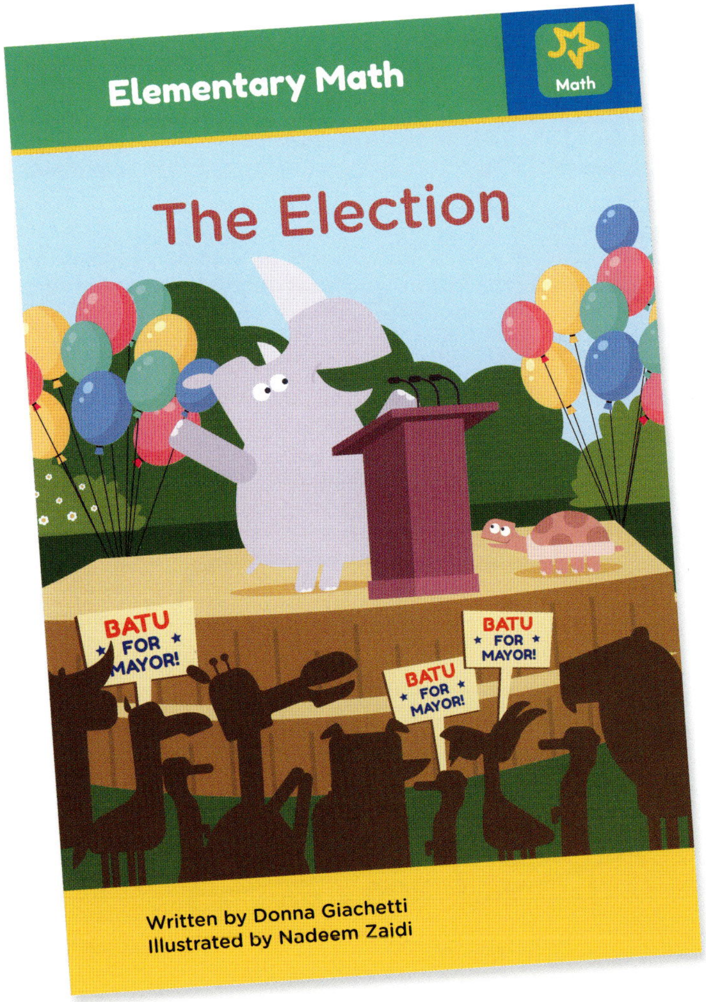

During reading:

Occasionally ask your child how they would write the number if there was one more or one less vote for a pal. Or ask your child to find the differences between the vote tallies, such as 10 and 100, or 100 and 1,000.

After reading:

Ask your child why Batu wanted to run for Mayor. What did they think of his reason? Which pal would your child vote for?

Allow family members to "vote" for a simple decision, such as the choice between two options for dinner. Your child can even make paper ballots and tally the votes.

Celebrate your child's progress by placing a star sticker in the workbook.

Super Shopper

Kazoo is going shopping for party supplies, but he needs help with the list. Look at the way the number for each item is written. Rewrite the number on the line next to each picture using digits.

one thousand eleven  ________

1,000 + 300 + 20 + 4  ________

  ________

  ________

Launch

Skip Counting Review

Two by Two

Remember, *skip counting* means to count forward by a number that is not one. To skip count, add the same number each time to the number that comes before it.

To skip count by 2s, you add 2 to get the next number.

You can start from any number and stop on any number. A hundred chart can help you see patterns.

Skip count from **41** to **61** by 2s. Notice that the digits in the ones place repeat the same pattern: **1**, **3**, **5**, **7**, **9**.

+2 +2 +2 +2

1	2	3	4	5	6	7	8	9	10
11	12	13	14	15	16	17	18	19	20
21	22	23	24	25	26	27	28	29	30
31	32	33	34	35	36	37	38	39	40
41	42	43	44	45	46	47	48	49	50
51	52	53	54	55	56	57	58	59	60
61	62	63	64	65	66	67	68	69	70
71	72	73	74	75	76	77	78	79	80
81	82	83	84	85	86	87	88	89	90
91	92	93	94	95	96	97	98	99	100

What's Next?

Skip count by 2s and write the numbers. Use the hundred chart to help you.

+2 +2 +2 +2 +2 +2

73 ____ ____ ____ ____ ____ **85**

+2 +2 +2 +2 +2 +2 +2

24 ____ ____ ____ ____ ____ ____ **38**

+2 +2 +2 +2 +2 +2

11 ____ ____ ____ ____ ____ **23**

+2 +2 +2 +2 +2 +2

82 ____ ____ ____ ____ ____ **94**

If your child needs more practice with skip counting, they can review Step 9, Lesson 2 in Math Fundamentals.

Skip Count Connect the Dots

Starting at 5, skip count by 2s to connect the dots and reveal the picture. Color it!

Hint: Some numbers in the picture are not part of the skip counting pattern.

Launch

Counting to 1,000

Seeing Patterns

A *pattern* is a sequence of repeating objects, shapes, or numbers that follows a rule.

You can use patterns to count to 1,000. You can count by 1s, or skip count by 2s, 5s, 100s, or with other numbers less than 1,000. Remember, if you can count to 100, you can use that same pattern to count up to 1,000.

Count by 1s	**991, 992, 993, 994, 995, 996, 997, 998, 999, 1,000**
Skip count by 5s	**975, 980, 985, 990, 995, 1,000**
Skip count by 100s	**100, 200, 300, 400, 500, 600, 700, 800, 900, 1,000**

Try It!

Match each number pattern, or sequence, with its rule.

950, 960, 970, 980, 990, 1,000
900, 925, 950, 975, 1,000, 1,025
990, 992, 994, 996, 998, 1,000

Skip counting by 2s
Skip counting by 10s
Skip counting by 25s

If your child is curious, have them try counting aloud to 1,000. Remind them that it's okay to take their time.

Race to 1,000

Each path counts up to 1,000 but starts on a different number. Fill in the missing numbers.
Circle the path that takes the fewest steps to reach 1,000.

Launch

Counting Patterns

Pattern Place

You can see patterns as part of a number sequence.
By identifying the pattern, you know what number comes next.

In this set of numbers, the pattern increases by 1 each time.

521	522	523	524	525	526

In this set of numbers, the pattern is skip counting by 10s.

320	330	340	350	360	370

Try It!

Find the pattern. Then complete the grids by filling in the missing numbers.
The first one has been done for you.

831	832		834		836

What's the pattern? ______

221	231		251	261	

What's the pattern? ______

400		600	700		900

What's the pattern? ______

Continue the Count

Figure out the pattern in each table, then fill in the missing numbers.

501	502			505					510
		513				517			

What's the pattern? ______

100			400				800		1,000

What's the pattern? ______

305	310	315			330				350
	360			375				395	
405									450

What's the pattern? ______

Greater Than, Less Than, Equal To

Math uses numbers with symbols called *signs* in place of words to write sentences that are read from left to right. Some signs compare numbers.

When comparing numbers, start by comparing the digits in the place with the greatest value. If the values are the same, compare the digits with the next greatest value. Continue until you find the answer.

$>$	greater than
$<$	less than
$=$	equal to, or the same

Which is greater, 2,691 or 2,689?
Use a place value chart to compare.

The **2**s in the thousands place are equal.
The **6**s in the hundreds place are equal.
But the **9** in the tens place is greater than the **8** in the tens place, because $90 > 80$.

2,691 $>$ 2,689

Thousands	Hundreds	Tens	Ones
2,	6	9	1

Thousands	Hundreds	Tens	Ones
2,	6	8	9

Try It!

Compare the two numbers below. Write each number in a place value chart. Then write the correct symbol in the circle.

6,467 ◯ 6,721

Thousands	Hundreds	Tens	Ones

Thousands	Hundreds	Tens	Ones

A Dicey Game

- Roll three dice and write the numbers in the boxes under Roll 1 in any order.
- Roll again and write the numbers in the boxes under Roll 2 in any order.
- In the circle, write the symbol that makes the sentence true.
- Keep rolling until you fill in all the boxes. Color the largest number you rolled **green**. Color the smallest number you rolled **blue**.

Roll 1		Roll 2
☐☐☐	○	☐☐☐

Roll 1		Roll 2
☐☐☐	○	☐☐☐

Roll 1		Roll 2
☐☐☐	○	☐☐☐

Roll 1		Roll 2
☐☐☐	○	☐☐☐

Roll 1		Roll 2
☐☐☐	○	☐☐☐

Roll 1		Roll 2
☐☐☐	○	☐☐☐

Materials needed: set of 3 dice and blue and green crayons or colored pencils

Launch  One More, Ten More, One Hundred More

In math, numbers can be compared by looking at the differences between the values of their digits. When adding or subtracting, it is important to add or subtract in the correct place.

Let's look at an example. Let's see how the number **238** changes when you add or subtract 1, 10, and 100 to the number.

1 less	Number	1 more
237	238	239

Think of 1 more or 1 less on a number line.

10 less	Number	10 more
228	238	248

Think of 10 more or 10 less on a section of a hundred chart.

221	222	223	224	225	226	227	228	229	230
231	232	233	234	235	236	237	238	239	240
241	242	243	244	245	246	247	248	249	250

100 less	Number	100 more
138	238	338

Think of 100 more or 100 less in vertical addition or subtraction problems.

$$\begin{array}{r} 238 \\ -\,100 \\ \hline 138 \end{array} \qquad \begin{array}{r} 238 \\ +\,100 \\ \hline 338 \end{array}$$

At a Crossroads

Look at the number in the center square of each grid. Follow the directions to write the numbers in the squares around it. The first one has been done for you.

	10 less 45	
1 less 54	55	1 more 56
	10 more 65	

	10 less	
1 less	163	1 more
	10 more	

	10 less	
1 less	284	1 more
	10 more	

More or Less

For each row, write 1, 10, or 100 more or less than the given number. The first one has been done for you.

100 less	10 less	1 less	Number	1 more	10 more	100 more
792	882	891	892	893	902	992
			746			
			391			

Play — The Greatest

Comparing Contest

Each starting bracket has 2 lines. Compare the numbers on each line and write the symbol to make the sentence true. Then write the 2 largest numbers in the next bracket and compare them. Continue until you are comparing the final two numbers. Write the greatest number on the next page. The first bracket has been done for you.

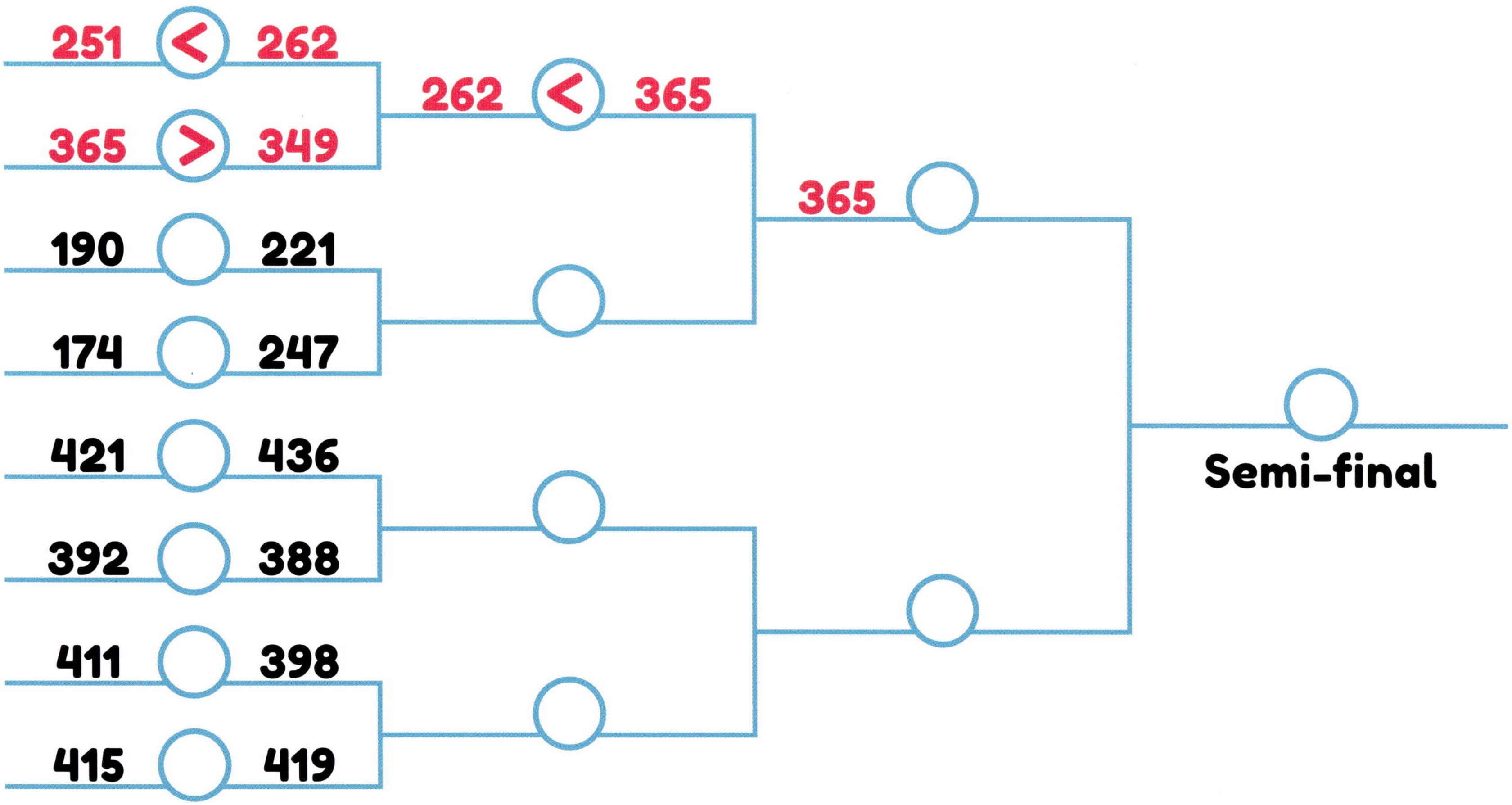

> greater than < less than = equal to, or the same

The greatest number is

_______.

242	◯	76
333	◯	323
142	◯	124
156	◯	165
309	◯	305
435	◯	431
299	◯	300
401	◯	399

Semi-final

Final

Read

The Flying Disc

Your child is ready to read *The Flying Disc*, a fantasy story about friendly aliens, the aliens' new toy, and a fictional device called a *spinometer* that measures speed. Throughout the story, your child will see number sentences showing various number comparisons.

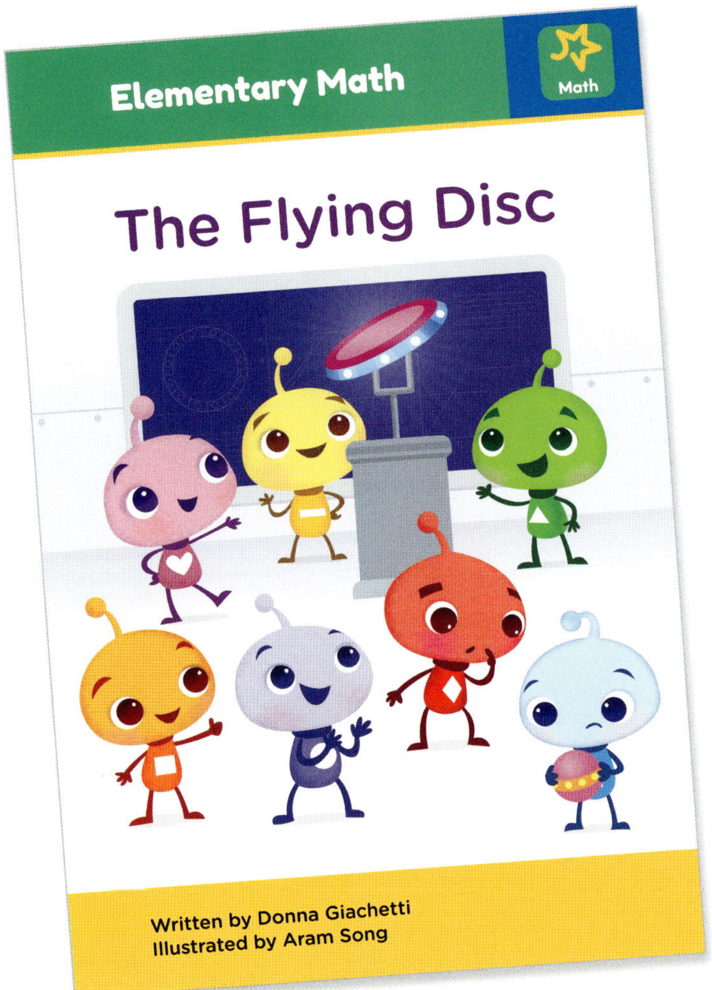

During reading:

Ask your child why the aliens travel to different planets to test their disc. Do the numbers increase or decrease as they go along?

Occasionally, cover up the symbol between the two numbers at the bottom of the page and ask your child if the numbers are less than, greater than, or equal to each other.

After reading:

What does Gluf want to test instead of a flying disc? Ask your child what number might be on the spinometer after they test it. Can your child think of any toy that might be even faster?

Celebrate your child's progress by placing a star sticker in the workbook.

Celebration Invitation

Gabby is having a party. She's inviting her bird pals but needs help with the invitations. Look at the number provided. Then write each street address number according to the directions. The first one has been done for you.

Number: 486

Directions	
100 Less	386 Oak Lane
10 More	_____ Elm Drive
1 More	_____ Maple Lane
10 Less	_____ Pine Street
100 More	_____ Ash Road
1 Less	_____ Spruce Street

Your child has completed Step 8!
Celebrate by placing a star sticker in the workbook.

Hooked on Math

It's Time to Celebrate!

____

has completed Elementary Math Steps 7 & 8
and is now ready to move on to the
Elementary Math Step 9 Workbook!

7

8

9

SKILLS MASTERED

- Understanding place value to 1,000
- Counting to 1,000
- Identifying counting patterns
- Comparing three- and four-digit numbers

Don't forget to share and tag on social media to celebrate! #hookedonmath

Credits

Design Team
Julie Linden Frazier
Lindsey Frazier

Editorial Team
Donna Giachetti
Cynthia Stierle

Illustrations
Dubit Limited

Educational Advisor
Lesley Freed, Curriculum Specialist

Published by Sandviks HOP, Inc.

Printed in China

Hooked on Math

Elementary Math

18043 (01/23)

H&
Co.

hookedandcompany.com

Hooked on Math

Elementary Math

Step 9

Workbook

Tips for Success

- Elementary Math focuses on understanding the importance of place value while working with numbers up to 1,000. It also introduces the key concept of regrouping and builds on the addition and subtraction skills introduced in Math Fundamentals.
- Work at a pace that's right for you and your child.
- Fit your session length to your child's attention span. It's okay to work with the App one day and this workbook the next.
- Before beginning, flip through the workbook and note any materials you might need. You should plan to have pencils, a pair of dice, and crayons ready.
- You'll see parent tips throughout this workbook. These tips are designed to answer questions you or your child might have about the learning.
- Repetition and review are essential to basic math fluency. Each workbook step builds on the ones before, so make sure your child has mastered the current lesson before moving forward.
- As you go through the workbook, encourage your child to explore the connection between numbers and the real world.
- Have fun with it!

Hooked on Math

Elementary Math

Step 9

Workbook

This workbook and the included storybooks introduce beginning math skills.

Step 9: Addition and Subtraction within 1,000

- Addition with regrouping
- Addition with regrouping word problems
- Subtraction with regrouping
- Subtraction with regrouping word problems

Hooked & Company

Contents

Lesson Overview

Welcome! This workbook provides supplemental activities that will solidify and expand the learning your child is enjoying with the Hooked on Math App. Follow the easy, step-by-step instructions to guide your child's exploration of numbers and operations. Every lesson helps your child build a strong foundation in elementary math skills while having fun!

1. Launch

Your child will be introduced to math concepts and skills through interactive lessons in the App on your mobile device or Windows or Mac computer. Download the App at:
hookedonmath.com/get-the-app

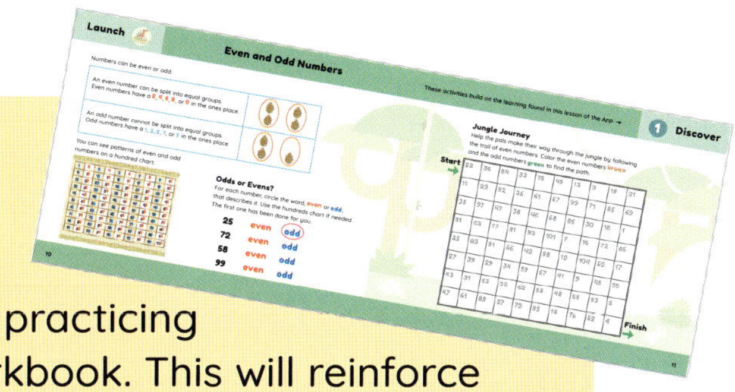

2. Discover

Next, your child will have fun exploring math concepts and practicing new skills in the workbook. This will reinforce the App lessons and provide a tangible application of the skills learned.

3. Play

In addition to the workbook activities, your child will enjoy two math-based storybooks that will strengthen their learning and add to the fun!

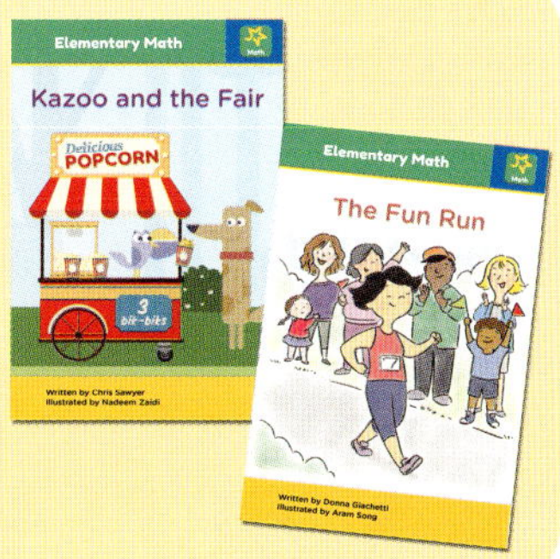

4. Celebrate

Celebrate your child's progress by placing a sticker in the workbook as learning milestones are reached. This also provides a helpful marker to remember where to begin next time.

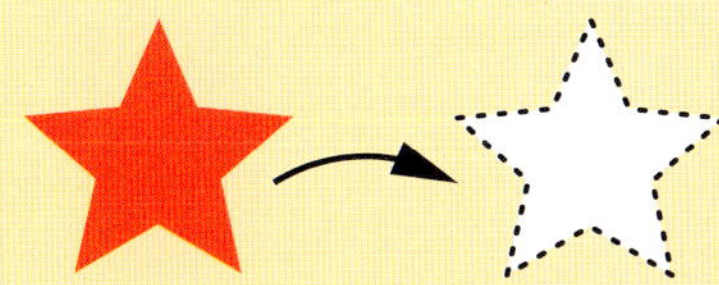

Special notes are included in this space to provide additional context related to what is being taught.

Place Value Sums to 1,000

Sum It Up!

Add the numbers and write the sum in the place value chart. Don't forget the comma!

Thousands	Hundreds	Tens	Ones

6,000 + 400 + 30 + 6

Thousands	Hundreds	Tens	Ones

8,000 + 0 + 20 + 5

Thousands	Hundreds	Tens	Ones

9,000 + 800 + 70 + 1

Sum-thing Is Missing

Look at the sum in each addition sentence. Write the value of the missing number. The first one has been done for you.

5,000 + 400 + 60 + 2 = 5,462

4,000 + 300 + ____ + 7 = 4,387

____ + 900 + 80 + 7 = 9,987

What's It Worth?

Read each word problem below. Then:

- Circle the number you are working with and underline the value you are trying to find.
- Write the number in the place value chart.
- Solve the problem and write the answer to the question. The first one has been done for you.

Read It	Write It	Answer it
The pals are planning a trip. They will travel 3,627 miles. What is the value of 3 in this number?	Thousands: 3, / Hundreds: 6 / Tens: 2 / Ones: 7	The value is 3,000.
Chow has a camera. He takes 1,952 pictures. What is the value of 9 in this number?	Thousands / Hundreds / Tens / Ones	The value is ______.
Together, the pals' backpacks weigh 1,649 pounds. Batu can lift them all! What is the value of 4 in this number?	Thousands / Hundreds / Tens / Ones	The value is ______.
Kazoo takes 7,648 steps on a hike. What is the value of 8 in this number?	Thousands / Hundreds / Tens / Ones	The value is ______.

Hop Around

Find the pattern in each table, then fill in the missing numbers.

50	100		200	
300	350		450	

The pattern is skip counting by _____.

600	620		660	680
	720	740		

The pattern is skip counting by _____.

705	720	735			780	795
	825		855		885	
915		945		975		

The pattern is skip counting by _____.

Going Up and Down

For each row, write 1, 10, or 100 more or less than the given number. The first one has been done for you.

100 less	10 less	1 less	Number	1 more	10 more	100 more
725	815	824	**825**	826	835	925
			850			
			875			
			900			
			925			

BONUS: What's the skip counting pattern in the *Number* column? ______

Celebrate finishing the review with a star sticker.

Double Check: Addition with Regrouping

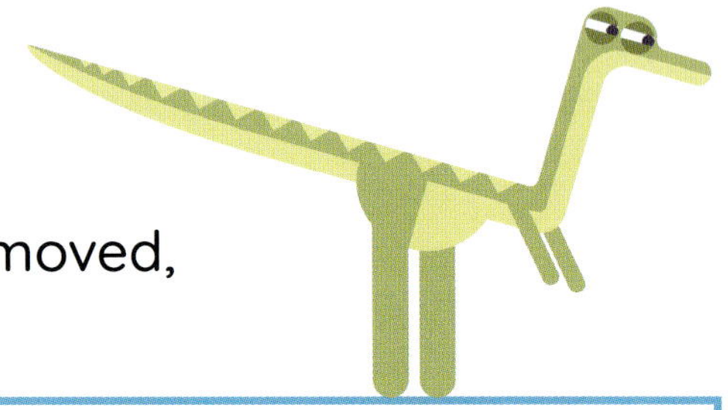

Let's review how to regroup when adding.

- You already know that each place can only have one digit, 0 – 9.
- If the sum of the digits in a place is ten or more, part of that sum must be moved, or *regrouped*, into the next place.

Two-digit numbers	1 26 + 39 65	The sum of the digits in the ones place is **15**, which is regrouped as 1 ten and 5 ones. The 1 ten is added to the tens place.
Two- and three-digit numbers	1 251 + 72 323	The sum of the digits in the tens place is **12**, which is regrouped as 2 tens and 1 hundred. The 1 hundred is added to the hundreds place.
Three-digit numbers	1 336 + 821 1,157	The sum of the digits in the hundreds place is **11**, which is regrouped as 1 hundred and 1 thousand. The 1 thousand is written in the thousands place. Don't forget the comma!

Carry On

Sometimes more than one place will need to be regrouped. Simply keep regrouping each place as you work from right to left.

```
 1 11
  369
+ 842
-----
1,211
```

Try It!

```
  682        278
+ 538      + 193
-----      -----
```

These activities build on the learning found in this lesson of the App. →

1 Discover

Rock Collection

Solve the addition problems in the rocks below. Match each sum to a number on the Color Key and then color the rock. Circle the rock with the greatest sum.

Color Key

93 = purple	251 = gray
739 = red	830 = yellow
953 = dark blue	1,003 = black
81 = orange	990 = green

$$\begin{array}{r} 65 \\ +\ 28 \\ \hline \end{array} \qquad \begin{array}{r} 196 \\ +\ 55 \\ \hline \end{array} \qquad \begin{array}{r} 248 \\ +\ 705 \\ \hline \end{array} \qquad \begin{array}{r} 641 \\ +\ 98 \\ \hline \end{array}$$

$$\begin{array}{r} 39 \\ +\ 42 \\ \hline \end{array} \qquad \begin{array}{r} 409 \\ +\ 421 \\ \hline \end{array} \qquad \begin{array}{r} 836 \\ +\ 154 \\ \hline \end{array} \qquad \begin{array}{r} 572 \\ +\ 431 \\ \hline \end{array}$$

Materials needed: crayons or colored pencils in purple, red, dark blue, orange, gray, yellow, black, and green

Snack Snatcher

Who Ate the Cookies from the Cookie Jar?

Help Detective Kazoo solve the mystery of the missing cookies. First, find each sum. Then draw a line to match the sum to a suspect to see who is full of cookies! The first one has been done for you.

Batu

$$\begin{array}{r} {\scriptstyle 1\,1\;\;} \\ 458 \\ +\,392 \\ \hline 850 \end{array}$$

Tali

$$\begin{array}{r} 622 \\ +\,281 \\ \hline \end{array}$$

Kip

$$\begin{array}{r} 442 \\ +\,262 \\ \hline \end{array}$$

Chow

$$\begin{array}{r} 395 \\ +\,395 \\ \hline \end{array}$$

Gabby

$$\begin{array}{r} 108 \\ +\,846 \\ \hline \end{array}$$

903
This pal only eats veggies.

850
This pal was taking a mud bath.

954
This pal was flying a plane.

704
This pal was delivering mail.

790
This pal is full of cookies!

Home, Sweet Home

Help the dinosaur find a path through the maze to reach its nest. Solve the problems. Then color the sums in the grid **green** to find the path to the nest.

508 + 113	153 + 562	292 + 64	785 + 89
897 + 259	226 + 185	140 + 399	279 + 643

Start →

715	621	356	725	327
366	875	1,156	933	489
611	569	411	539	1,256
1,146	891	415	874	922

Finish

Double Check: Addition Word Problems with Regrouping

Let's review the steps to solve addition word problems—and remember to regroup when you're adding!

Read the problem. Search for key information, such as words or phrases like *all together, join,* or *in all.* Circle or underline the information.
Set up a horizontal math number sentence. Don't forget math symbols like the plus sign and the equal sign.
Flip the problem to be vertical and solve it. Remember to regroup!
Use a complete sentence to answer the question the word problem is asking.

Batu walks (389) steps to his favorite mud puddle. He jumps in for a refreshing mud bath. Later, he takes the long way home and needs (448) muddy steps to get there. How many steps did he take all together to go to the mud puddle and back?

389 + 448 = ?

$$\begin{array}{r} {}^{1\,1} \\ 389 \\ +\,448 \\ \hline 837 \end{array}$$

Batu took **837** steps to get to the mud puddle and back.

Write the Story

Look at the problem below. Finish writing the story!

$$\begin{array}{r} {}^{1} \\ 837 \\ +\,749 \\ \hline 1{,}586 \end{array}$$

Batu wanted to get more exercise. . .

Story Problems

Read the story in each word problem below. Then:

- Circle the numbers you need to work with.
- Underline the words that tell what operation to perform.
- Write each number sentence horizontally, and then vertically.
- Solve the problem and write the answer to the question.

Story	Horizontal	Vertical	Answer
The pals planted 350 corn seeds in the east field. They planted 535 seeds in the west field. How many seeds did they plant in all?			
The pals used 436 gallons to water the corn in the east field. They used 575 gallons in the west field. How many gallons of water did they use all together?			
The corn grew well! At harvest time, they gathered 175 bushels of corn from the east field. They gathered 387 bushels from the west field. How many bushels did they gather all together?			

Encourage your child to use scratch paper if they need more room to write.

Where Is Chow?

Chow is going on vacation. Where is he flying to? Find out by solving the addition problems on this page. Then find and color the answers in the path on the opposite page to see where Chow is going on vacation! The first one has been done for you.

$$\begin{array}{r} 845 \\ +\,154 \\ \hline 999 \end{array} \quad \begin{array}{r} 382 \\ +\,219 \\ \hline \end{array} \quad \begin{array}{r} 527 \\ +\,354 \\ \hline \end{array} \quad \begin{array}{r} 193 \\ +\,471 \\ \hline \end{array} \quad \begin{array}{r} 221 \\ +\,279 \\ \hline \end{array}$$

$$\begin{array}{r} 724 \\ +\,187 \\ \hline \end{array} \quad \begin{array}{r} 482 \\ +\,375 \\ \hline \end{array} \quad \begin{array}{r} 347 \\ +\,589 \\ \hline \end{array} \quad \begin{array}{r} 744 \\ +\,199 \\ \hline \end{array} \quad \begin{array}{r} 221 \\ +\,538 \\ \hline \end{array}$$

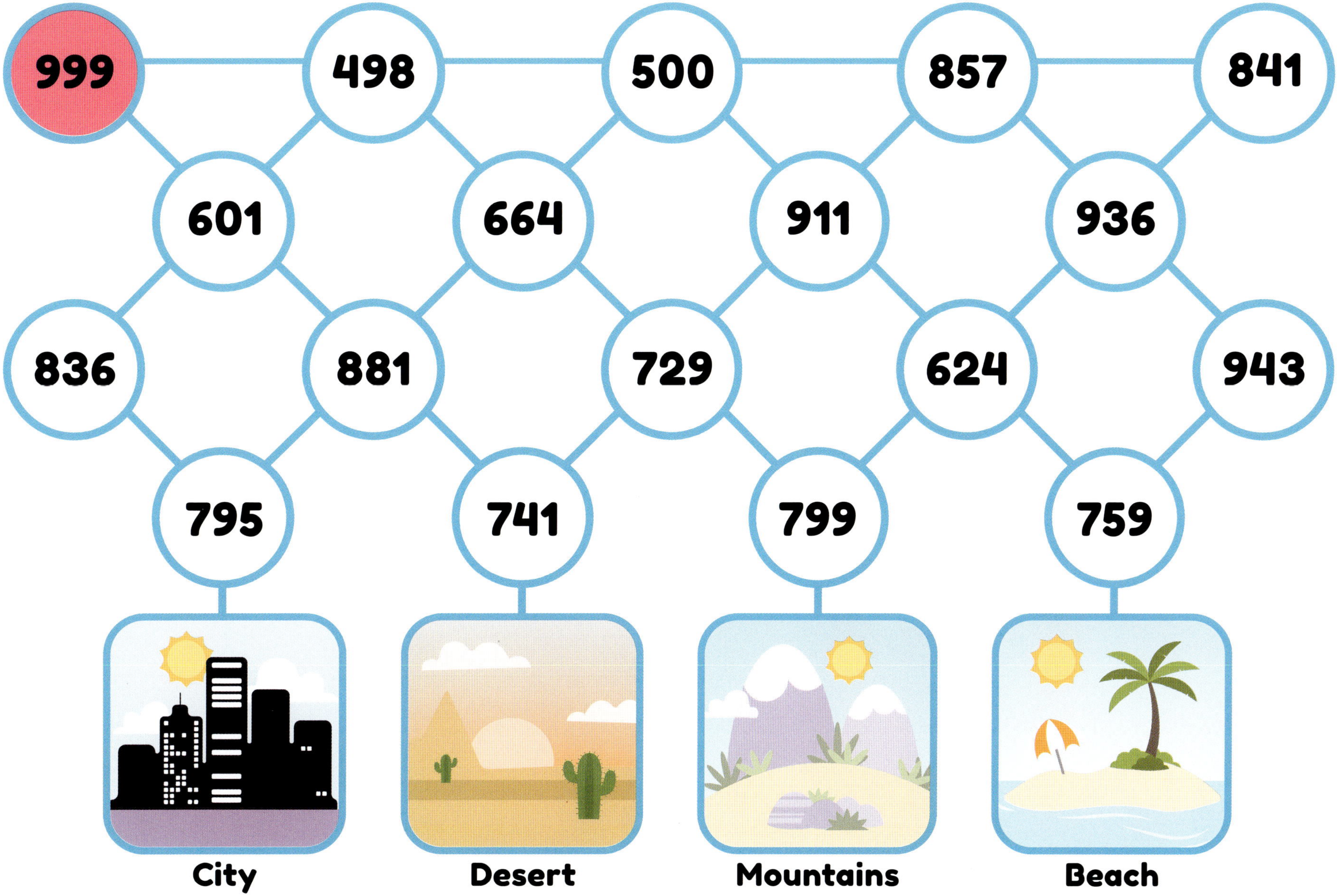
999
498
500
857
841
601
664
911
936
836
881
729
624
943
795
741
799
759
City
Desert
Mountains
Beach

Menu Mix Up

The pals went out to eat, but their orders got mixed up. Solve the addition problems next to each pal and next to each food. Then match each pal to the food that has the same answer. The first one has been done for you.

Batu

760
+ 44
804

Tali

568
+ 325

Kazoo

224
+ 194

Chow

491
+ 356

Gabby

677
+ 322

206
+ 793

256
+ 591

461
+ 432

365
+ 439
804

175
+ 243

Read

Your child is ready to read *Kazoo and the Fair*, a humorous story that features addition with regrouping up to 1,000. In this story, the characters are raising money and need to reach a goal of 1,000 *bik-biks*, a fictional type of money.

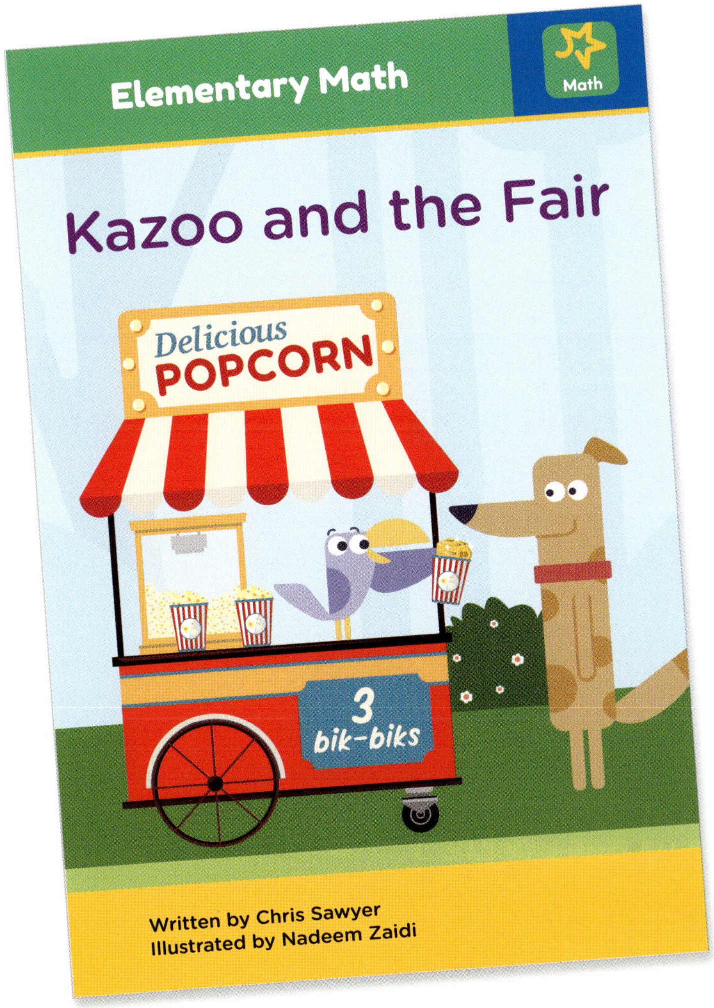

During reading:

The equations at the bottom of the pages are a running tally of the funds for the library, as each pal makes their contribution. Occasionally ask your child how close the pals are to reaching their goal. Can your child estimate how many more bik-biks the pals need? As a challenge, ask your child how they might use subtraction to determine exactly how many more bik-biks are needed.

After reading:

Has your child ever participated in a fundraiser, either by buying or selling an item? What did they do? What did they think of it? Ask your child if there is a cause for which they'd like to raise money. What is it, and why is it important?

Celebrate your child's progress by placing a star sticker in the workbook.

Launch

Double Check: Subtraction with Regrouping

Let's review how to regroup when subtracting.

- You already know that each place can only have one digit, 0 – 9.
- In each place, you cannot subtract a greater digit from a smaller digit. Instead, you must move, or regroup, an amount from the next place and add it to the smaller digit so that you can subtract.

Two-digit numbers	3 1 ~~4~~2 - 19 23	In the ones place, 9 cannot be subtracted from 2. The 4 tens are regrouped into **3** tens and 10 ones. The 10 ones are added to the **2** ones. Now **9** ones can be subtracted from 12 ones.
Two- and three-digit numbers	1 ~~2~~44 - 172 72	In the tens place, 7 cannot be subtracted from 4. The 2 hundreds are regrouped into **1** hundred and 10 tens. The 10 tens are added to the **4** tens. Now **7** tens can be subtracted from 14 tens.
Three-digit numbers	1 ~~1~~,836 - 921 915	In the hundreds place, 9 cannot be subtracted from 8. The **1** thousand is regrouped into 10 hundreds. No digits remain in the thousands place. The 10 hundreds are added to the **8** hundreds. Now **9** hundreds can be subtracted from 18 hundreds.

If your child wants to check an answer, explain that they can add the difference to the number they are subtracting. If they get the starting number, they've solved the problem correctly.

Subtraction Tic-Tac-Toe

Solve each subtraction problem to find the difference.

- Circle all the problems that have matching answers.
- Draw an **X** through the problems that don't have matching answers.
- Which problems win?

831 − 336	933 − 377	621 − 126
769 − 588	752 − 257	988 − 189
604 − 252	711 − 512	844 − 349

Borrow a Bit More

Sometimes more than one place will need to be regrouped. Simply keep regrouping each place as you work from right to left.

```
 641
 751
-578
----
 173
```

On Your Mark, Get Set . . .

The pals are having a race. Find the difference for each problem. Then follow the directions on the next page to see who wins the race.

Kip

703	804	907	779
- 165	- 385	- 427	- 486

Kazoo

985	614	259	608
- 491	- 359	- 106	- 499

Batu

532	302	528	945
- 478	- 174	- 113	- 685

GO!

Look at the differences on the previous page.

- Circle the differences that are odd numbers. Color 1 square for each odd number a pal has.
- Draw a box around the differences that are even numbers. Color 2 squares for each even number a pal has.
- Which pal wins?

Finish

Kip							
Kazoo							
Batu							

Double Check: Subtraction Word Problems with Regrouping

Let's review the steps to solve subtraction word problems—and remember to regroup!

Read the problem. Search for key information, such as words or phrases like *take away*, *left*, or *fewer*. Circle or underline the information.
Set up a horizontal math number sentence. Don't forget math symbols like the minus sign and the equal sign.
Flip the problem to be vertical and solve it. Remember to regroup!
Use a complete sentence to answer the question the word problem is asking.

Tali bought a roll of stickers. The roll had 375 stickers. She gave 136 stickers to her pals. How many stickers does she have left?

375 - 136 = ?

$$\begin{array}{r} \overset{6\,1}{3\not{7}5} \\ -\ 136 \\ \hline 239 \end{array}$$

Tali has **239** stickers left.

Write the Story

Look at the problem below. Finish writing the story!

$$\begin{array}{r} \overset{1}{\not{2}39} \\ -\ 187 \\ \hline 52 \end{array}$$

Tali had 239 stickers left. . .

Story Problems

Read the story in each word problem below. Then:

- Circle the numbers you need to work with.
- Underline the words that tell what operation to perform.
- Write each number sentence horizontally, and then vertically.
- Solve the problem and write the answer to the question.

Story	Horizontal	Vertical	Answer
Mr. Fox took 1,236 pictures with his new phone. But his paw covered the camera for 438 pictures. *Oops!* He deleted them. How many pictures were left?			
While he was at the beach, he took 673 pictures. Of those pictures, 135 were pictures of shells. How many pictures were *not* shells?			
Mr. Fox took 325 pictures of the sunset. He took 157 fewer pictures of the sunrise. How many pictures did he take of the sunrise?			

Fresh Fruit

The meerkats are picking different fruits to make smoothies for a big meerkat gathering. Which fruit is their favorite? To find out, solve the subtraction problems on this page. Then find and color the answers in the maze on the opposite page to see which fruit they like the most.

$$\begin{array}{r} 866 \\ -\ 274 \\ \hline \end{array} \qquad \begin{array}{r} 532 \\ -\ 439 \\ \hline \end{array} \qquad \begin{array}{r} 727 \\ -\ 144 \\ \hline \end{array} \qquad \begin{array}{r} 990 \\ -\ 278 \\ \hline \end{array} \qquad \begin{array}{r} 531 \\ -\ 528 \\ \hline \end{array}$$

$$\begin{array}{r} 724 \\ -\ 187 \\ \hline \end{array} \qquad \begin{array}{r} 822 \\ -\ 345 \\ \hline \end{array} \qquad \begin{array}{r} 349 \\ -\ 189 \\ \hline \end{array} \qquad \begin{array}{r} 607 \\ -\ 205 \\ \hline \end{array} \qquad \begin{array}{r} 486 \\ -\ 338 \\ \hline \end{array}$$

Just Desserts

Color each number that is an answer to a problem on the previous page. Then follow the path to see which fruit smoothie the meerkats like best.

592	97	626	271	655	
93	5	458	165	598	
583	537	477	160	438	
712	3	499	402	148	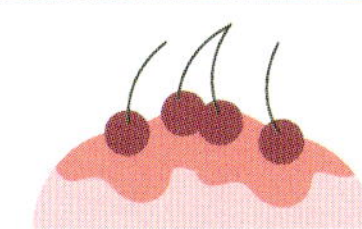

An Eggs-cellent Lesson

Gabby is teaching some little chicks how to count. Solve each subtraction problem to create the Code Key. Then use it to discover the answers to the riddles on the next page.

Code Key

756 - 161 = E	876 - 764 = G	592 - 336 = S	424 - 263 = A
542 - 438 = M	259 - 187 = P	730 - 616 = L	971 - 437 = T
419 - 341 = R	756 - 165 = C	179 - 84 = D	941 - 658 = I

Code Breaker

Use the Code Key on the previous page to solve these riddles.

Riddle: Why was Gabby so good at teaching math?

Answer: Because she showed the chicks plenty of

___ ___ ___ ___ - ___ ___ ___ ___ ___ ___

595 112 112 256 - 161 104 72 114 595 256

Riddle: How could the chicks earn bonus points in Gabby's class?

Answer: They could do

___ ___ ___ ___ - ___ ___ ___   ___ ___ ___ ___ ___ ___

595 112 112 256 - 534 78 161   591 78 595 95 283 534

Back and Forth

The pals are having an addition and subtraction tug-of-war contest.

- First, solve the problems for each team on both pages.
- Write both answers in the center column. Then use the correct symbol to show if the numbers are less than, greater than, or equal to each other.
- Place a check mark in the *Win* column of the team with the greater number. The first one has been done for you.

	Team Blue	Win		Win	Team Orange
Round 1	256 + 456 712	✓	712 > 446		735 - 289 446
Round 2	144 + 296				943 - 269
Round 3	416 + 487				701 - 65

Keep playing. Add, subtract, and compare to see which team wins the tug-of-war contest.

	Team Blue	Win		Win	Team Orange
Round 4	179 + 555				1,082 - 385
Round 5	411 + 271				945 - 185
Round 6	144 + 296				943 - 269
Round 7	416 + 487				701 - 65

Count each team's check marks. Write them here. Which team won?

Team Blue _____ **Team Orange** _____

Read — The Fun Run

Your child is ready to read *The Fun Run,* a sweet story about Miss Gallop, a runner who tries to run 1,000 paces and gains an unexpected companion along the way. Throughout the story, your child will see the number of steps Miss Gallop has left as she runs along.

During reading:

At the beginning of the story, we learn that Miss Gallop has never won a race. Ask your child if they think she will win this time. Why or why not? Is winning important, or is participating?

Occasionally before turning a page, ask your child how many paces they think Miss Gallop will run next. Then see if they can calculate how many paces she will have left to run based on that guess.

After reading:

What would your child do if they discovered a stray dog while running a race?

Ask your child if they think 1,000 paces is very far. Do they think they can run that far? The next time you and your child take a walk, consider using an app to track your steps. Point out when you reach 1,000!

Celebrate your child's progress by placing a star sticker in the workbook.

Flip the Switch

Fill in the missing number in each equation. The first one has been done for you.

$$\begin{array}{r} {\scriptstyle 11} \\ 195 \\ +\ \boxed{366} \\ \hline 561 \end{array} \qquad \begin{array}{r} {\scriptstyle 4\,5\,1} \\ \not{5}\not{6}1 \\ -\ 195 \\ \hline \boxed{366} \end{array}$$

$$\begin{array}{r} 871 \\ +\ 61 \\ \hline \_\_\_\_ \end{array} \qquad \begin{array}{r} 932 \\ -\ \_\_\_\_ \\ \hline 871 \end{array}$$

$$\begin{array}{r} 276 \\ +\ 445 \\ \hline \_\_\_\_ \end{array} \qquad \begin{array}{r} \_\_\_\_ \\ -\ 445 \\ \hline 276 \end{array}$$

$$\begin{array}{r} 289 \\ +\ 621 \\ \hline \_\_\_\_ \end{array} \qquad \begin{array}{r} 910 \\ -\ \_\_\_\_ \\ \hline 289 \end{array}$$

$$\begin{array}{r} 536 \\ +\ 148 \\ \hline \_\_\_\_ \end{array} \qquad \begin{array}{r} 684 \\ -\ \_\_\_\_ \\ \hline 148 \end{array}$$

$$\begin{array}{r} 355 \\ +\ \_\_\_\_ \\ \hline 775 \end{array} \qquad \begin{array}{r} 775 \\ -\ 355 \\ \hline \_\_\_\_ \end{array}$$

Your child has completed Step 9!
Celebrate by placing a star sticker in the workbook.

Hooked on Math

It's Time to Celebrate!

____

has completed Elementary Math Step 9
and is now ready to move on to
Core Math!

SKILLS MASTERED

- Addition with regrouping
- Addition with regrouping word problems
- Subtraction with regrouping
- Subtraction with regrouping word problems

Don't forget to share and tag on social media to celebrate! #hookedonmath

Credits

Design Team
Julie Linden Frazier
Lindsey Frazier

Editorial Team
Donna Giachetti
Cynthia Stierle

Illustrations
Dubit Limited

Educational Advisor
Lesley Freed, Curriculum Specialist

Published by Sandviks HOP, Inc.

Printed in China

Hooked on Math

Elementary Math

18046 (05/23)

H& Co.

hookedandcompany.com

Hooked on Math

Elementary Math

Steps 5 & 6

Workbook

Tips for Success

- Elementary Math focuses on understanding the importance of place value while working with numbers up to 1,000. It also introduces the key concept of regrouping and builds on the addition and subtraction skills introduced in Math Fundamentals.
- Work at a pace that's right for you and your child.
- Fit your session length to your child's attention span. It's okay to work with the App one day and this workbook the next.
- Before beginning, flip through the workbook and note any materials you might need. You should plan to have pencils, a pair of dice, and crayons ready.
- You'll see parent tips throughout this workbook. These tips are designed to answer questions you or your child might have about the learning.
- Repetition and review are essential to basic math fluency. Each workbook step builds on the ones before, so make sure your child has mastered the current lesson before moving forward.
- As you go through the workbook, encourage your child to explore the connection between numbers and the real world.
- Have fun with it!

Hooked on Math

Elementary Math

Steps 5 & 6

Workbook

This workbook and the included storybooks introduce elementary math skills.

Step 5: Subtraction to 100
- Subtracting on a number line
- Counting back on a hundred chart
- Skip counting back on a hundred chart

Step 6: Subtraction within 100
- Subtraction with regrouping
- Subtraction word problems with regrouping
- Place value blocks subtraction

Hooked & Company

Contents

Lesson Overview

Welcome! This workbook provides supplemental activities that will solidify and expand the learning your child is enjoying with the Hooked on Math App. Follow the easy, step-by-step instructions to guide your child's exploration of numbers and operations. Every lesson helps your child build a strong foundation in elementary math skills while having fun!

1. Launch

Your child will be introduced to math concepts and skills through interactive lessons in the App on your mobile device or Windows or Mac computer. Download the App at:

hookedonmath.com/get-the-app

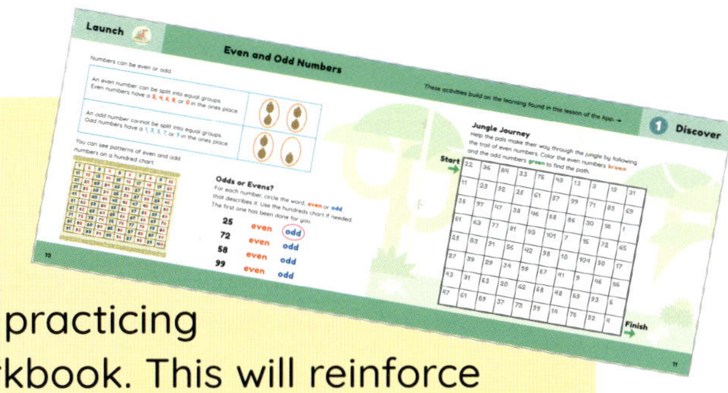

2. Discover

Next, your child will have fun exploring math concepts and practicing new skills in the workbook. This will reinforce the App lessons and provide a tangible application of the skills learned.

3. Play

In addition to the workbook activities, your child will enjoy two math-based storybooks that will strengthen their learning and add to the fun!

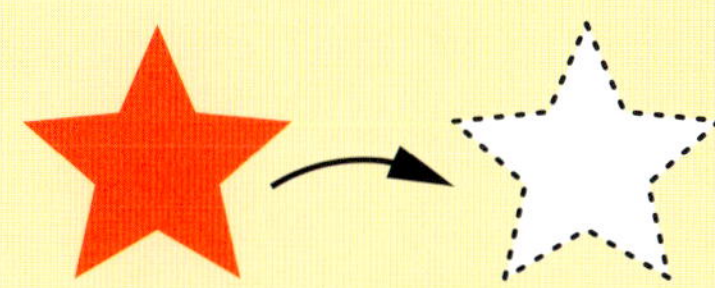

4. Celebrate

Celebrate your child's progress by placing a sticker in the workbook as learning milestones are reached. This also provides a helpful marker to remember where to begin next time.

Special notes are included in this space to provide additional context related to what is being taught.

Tens and Ones Word Problems

In a Few Words

- Read each story. Circle the numbers in the problem, and underline the key words or phrases that tell you which operation to perform—addition or subtraction.
- Circle the horizontal number sentence that matches the story.
- Flip the horizontal number sentence to a vertical number sentence and solve it.
- Write the answer to the question in sentence form. The first one has been done for you.

Story	Horizontal	Vertical	Answer
It's clean-up day! Kazoo collected (26) old cans. Tali picked up (42) empty bottles. How many bottles and cans did they gather <u>in all</u>?	(26 + 42 = ?) 42 - 26 = ? 46 + 22 = ?	26 + 42 68	They gathered 68 bottles and cans in all.
Zola dropped a jar filled with 96 marbles. Her friends helped her find 72 of them. How many marbles did she lose?	96 + 72 = ? 72 - 69 = ? 96 - 72 = ?		
Chow's team scored 78 points. Batu's team had 66 points. How many more points did Chow's team score than Batu's team?	87 - 66 = ? 78 - 66 = ? 78 + 66 = ?		

How Many?

Look at the place value flats, rods, and blocks. Write how many each represents on the line below it. Then add the numbers together and write the sum. The first one has been done for you.

500 + 10 + 6 = 516

____ + ____ + ____ = ____

____ + ____ + ____ = ____

Up and Away!

For each problem, add the numbers and write the sum in the shaded area. Then, on the next page, fill in the area with the matching number with the same color. The first one has been done for you.

Tens	Ones
1	
4	4
+ 3	8
8	2

Tens	Ones
5	5
+ 2	6

Tens	Ones
5	9
+ 1	4

Tens	Ones
4	5
+ 1	8

Tens	Ones
7	7
+ 1	7

Tens	Ones
4	7
+ 2	5

Tens	Ones
1	6
+ 1	9

Tens	Ones
2	4
+ 6	7

Look Up!
Color the picture according to the code on the previous page.

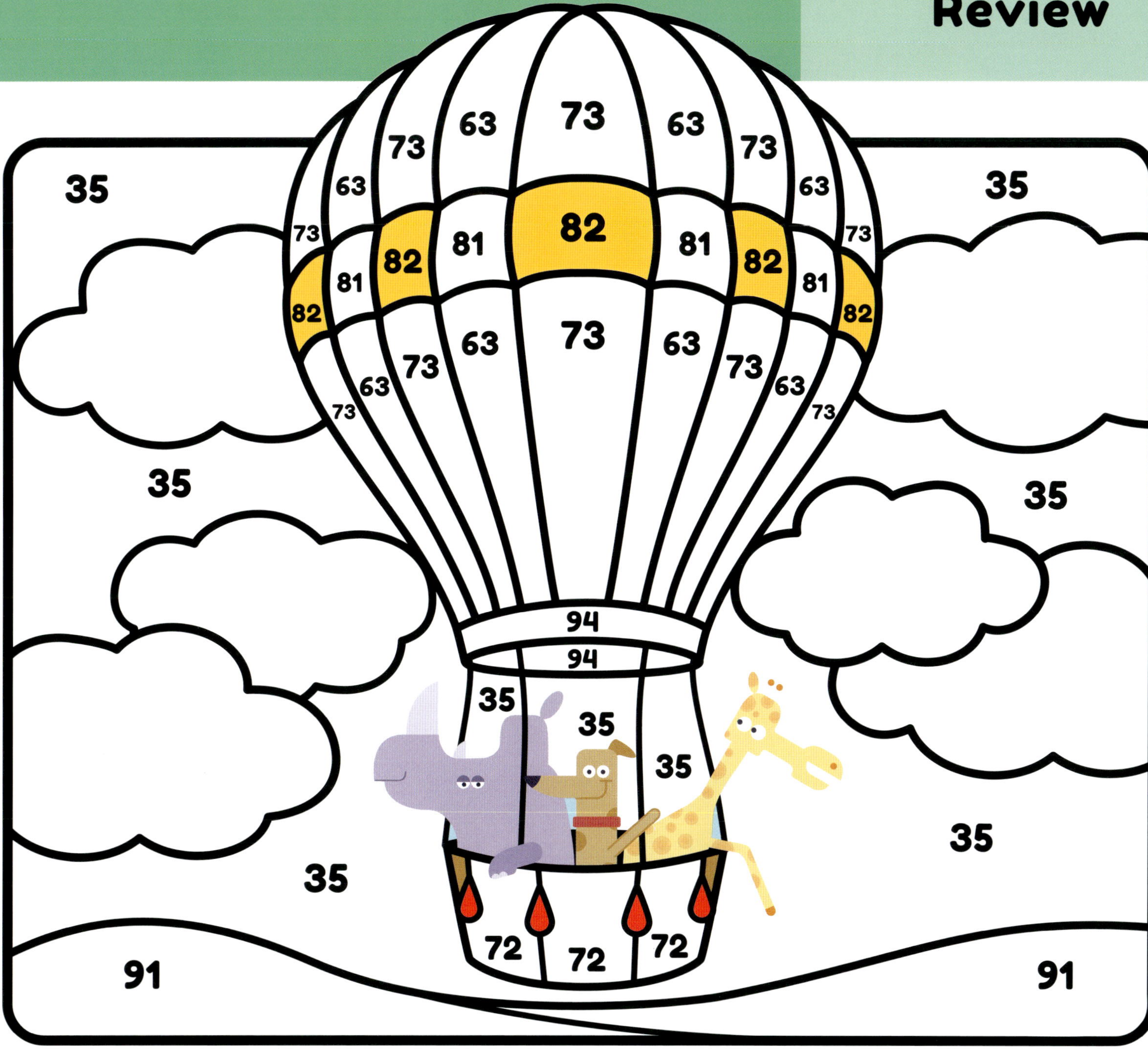

Materials needed: crayons or colored pencils in purple, light blue, dark blue, orange, red, brown, and green

That's How I Roll

- Roll a pair of dice to create a double-digit number. Write it in Roll 1.
- Roll again to create another double-digit number. Write it in Roll 2.
- Color the matching place value blocks for both rolls.
- Add the rolls together to find the sum, regrouping if needed.
  Then color the place value flats, rods, and blocks to show the sum.
- Keep rolling until all the boxes are filled in.

	Hundreds	Tens	Ones
		1	
Roll 1		6	6
Roll 2	+	4	5
SUM	1	1	1

	Hundreds	Tens	Ones
Roll 1			
Roll 2	+		
SUM			

	Hundreds	Tens	Ones
Roll 1			
Roll 2	+		
SUM			

	Hundreds	Tens	Ones
Roll 1			
Roll 2	+		
SUM			

Materials needed: Pair of dice

Roll 1 Roll 2 SUM

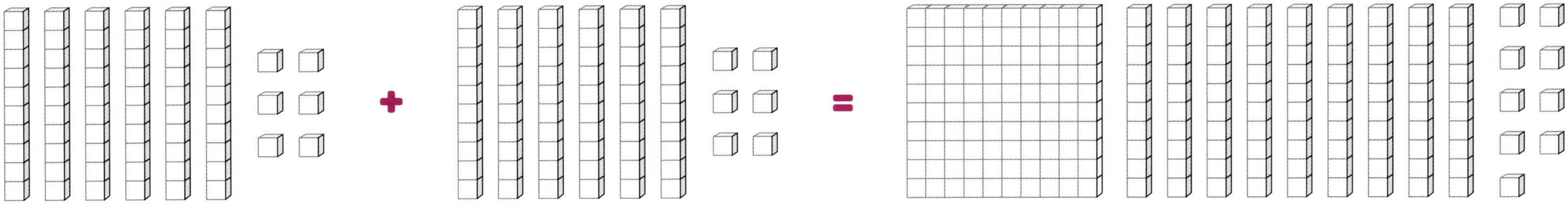

Roll 1 Roll 2 SUM

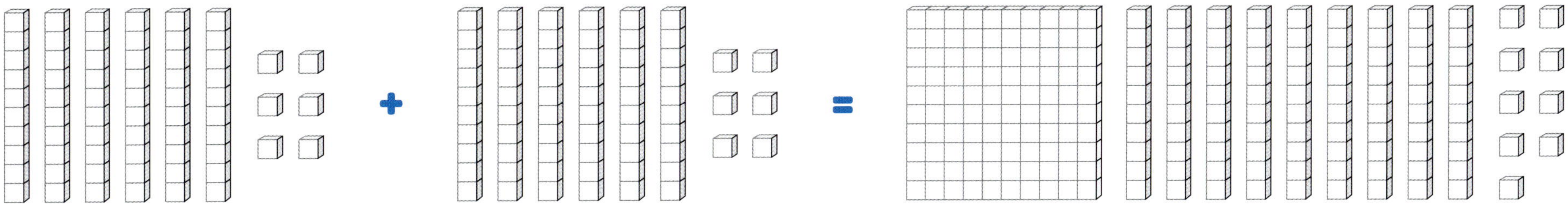

Roll 1 Roll 2 SUM

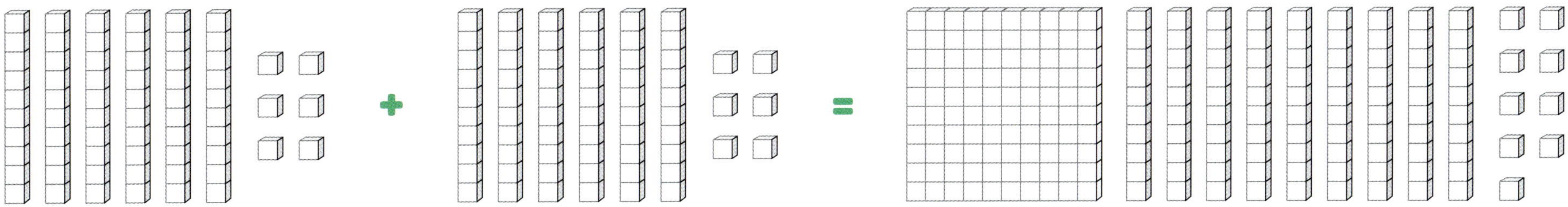

Celebrate finishing the review with a star sticker.

Subtracting on a Number Line

Remember, when the number being subtracted is small, you can use a number line along with the counting back strategy to find the difference. But what if the number being subtracted is large?

If the two numbers are close together, find the distance between the two numbers instead. This is called *subtracting down* on a number line.

Let's look at an example:

94 - 86 = ?

Mark both numbers on the number line. Then count the distance, or jumps, between the two numbers.	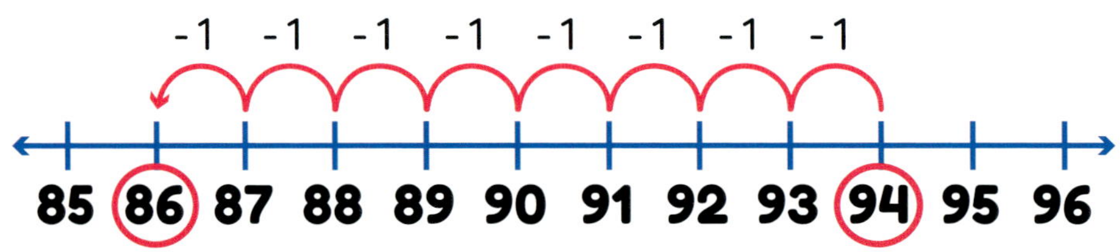
There are 8 jumps between 94 and 86.	**94 - 86 = 8**

Hop, Skip, and a Jump

This strategy is best used with two numbers that are close together. Look at the problems below. Circle the ones with two numbers that are close together. Draw an **X** through the ones that are farther apart.

83 - 75 = 8   **66 - 18 = 48**   **32 - 27 = 5**   **48 - 19 = 29**

A number line is a useful tool for modeling strategies in addition and subtraction, helping your child visualize the action of the operation as well as the distance between numbers.

Short Distance

Use the number line to solve the subtraction problems below. First circle the numbers in the problem on the number line. Then draw the jumps between the numbers. Finally, count the jumps and write the difference. The first one has been done for you.

77 - 69 = 8

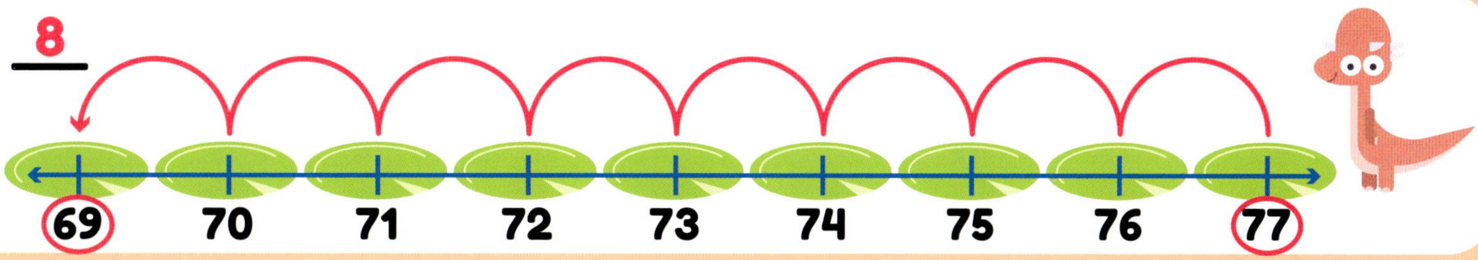

52 - 47 = ___

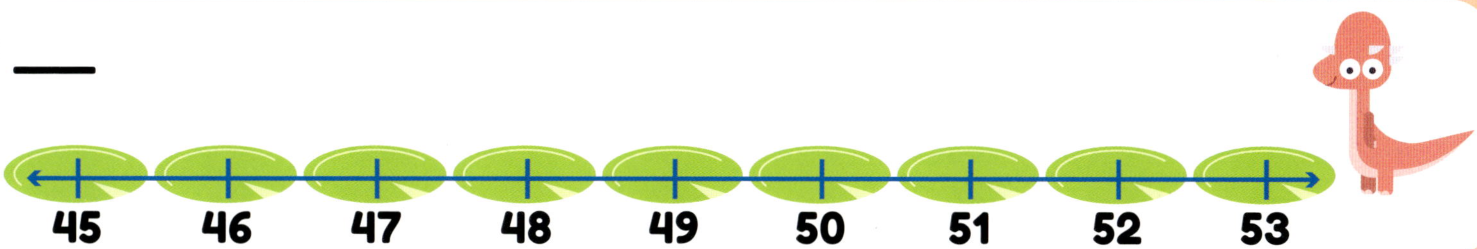

36 - 29 = ___

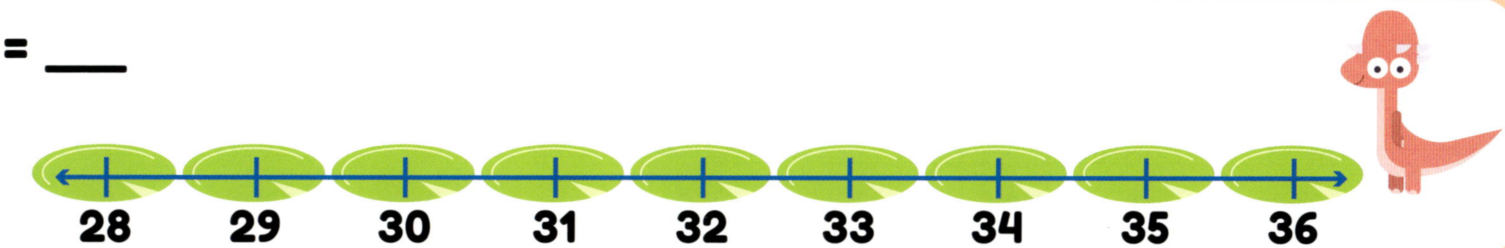

Launch

Count Back to Subtract Ones on a Hundred Chart

The counting back strategy involves counting numbers in backward order, starting from a given number. A hundred chart can help.

Count Back by Ones

When you are subtracting an amount less than 10, you can count back by ones.

Look at the problem 88 - 6 = ?

Start at **88**. Count back 6 ones: **87**, **86**, **85**, **84**, **83**, **82**.

88 - 6 = **82**

71	72	73	74	75	76	77	78	79	80
81	82	83	84	85	86	87	88	89	90
91	92	93	94	95	96	97	98	99	100

Try another! 53 - 7 = ?

Start at **53**. Count back 7 ones: **52**, **51**, **50**, **49**, **48**, **47**, **46**.

53 - 7 = **46**

31	32	33	34	35	36	37	38	39	40
41	42	43	44	45	46	47	48	49	50
51	52	53	54	55	56	57	58	59	60

Easy Does It!

For each problem, circle the starting number on the hundred chart. Then color in the number of ones you need to count back. Circle and write the answer.

1	2	3	4	5	6	7	8	9	10
11	12	13	14	15	16	17	18	19	20
21	22	23	24	25	26	27	28	29	30
31	32	33	34	35	36	37	38	39	40
41	42	43	44	45	46	47	48	49	50

47 - 8 = ___

51	52	53	54	55	56	57	58	59	60
61	62	63	64	65	66	67	68	69	70
71	72	73	74	75	76	77	78	79	80
81	82	83	84	85	86	87	88	89	90
91	92	93	94	95	96	97	98	99	100

87 - 9 = ___

What's the Problem?

Each hundred chart uses the counting back strategy to find the difference.

- Write the starting number.
- Count the ones and write the number being taken away.
- Write the difference. The first one has been done for you.

1	2	3	4	5	6	7	8	9	10
11	12	13	14	15	16	17	18	19	20
21	22	23	24	25	26	27	28	29	30
31	32	33	34	35	36	37	38	39	40
41	42	43	44	45	46	47	48	49	50

39 - 4 = 35

51	52	53	54	55	56	57	58	59	60
61	62	63	64	65	66	67	68	69	70
71	72	73	74	75	76	77	78	79	80
81	82	83	84	85	86	87	88	89	90
91	92	93	94	95	96	97	98	99	100

___ - ___ = ___

1	2	3	4	5	6	7	8	9	10
11	12	13	14	15	16	17	18	19	20
21	22	23	24	25	26	27	28	29	30
31	32	33	34	35	36	37	38	39	40
41	42	43	44	45	46	47	48	49	50

___ - ___ = ___

1	2	3	4	5	6	7	8	9	10
11	12	13	14	15	16	17	18	19	20
21	22	23	24	25	26	27	28	29	30
31	32	33	34	35	36	37	38	39	40
41	42	43	44	45	46	47	48	49	50

___ - ___ = ___

1	2	3	4	5	6	7	8	9	10
11	12	13	14	15	16	17	18	19	20
21	22	23	24	25	26	27	28	29	30
31	32	33	34	35	36	37	38	39	40
41	42	43	44	45	46	47	48	49	50

___ - ___ = ___

51	52	53	54	55	56	57	58	59	60
61	62	63	64	65	66	67	68	69	70
71	72	73	74	75	76	77	78	79	80
81	82	83	84	85	86	87	88	89	90
91	92	93	94	95	96	97	98	99	100

___ - ___ = ___

Visualizing counting back on a hundred chart can help your child with the concept of subtraction with and without regrouping.

New Problem? No Problem!

Mental math can help you solve problems that you haven't seen before by using what you know. For example, how would you solve the problem 12 - 6 = 13 - ?

Notice the equal sign between the two equations. That means the answer to both sentences is the same.	**12 - 6 = 13 - ?**
Notice that the second number sentence has a **?**, or an unknown. That is what you are solving for.	**12 - 6 = 13 - ?**
The doubles fact 6 + 6 = 12 helps you solve for **12 - 6 = 6**. That means the answer to the second number sentence must also be **6**. You can put the subtraction sentence on the other side of the equal sign.	**12 - 6 = 13 - ?** **6 = 13 - ?** **13 - ? = 6**

Let's use fact families to find the missing number. Which number is missing?

13

6 +, - **7**

6 + ___ = 13

___ + 6 = 13

13 - 6 = ___

13 - ___ = 6

It's **7**!
That means the unknown in the equation is **7**.

12 - 6 = 13 - ?
12 - 6 = 13 - 7
6 = 6

This lesson focuses on mathematical reasoning skills. Whenever your child encounters an unfamiliar problem, encourage them to break it into smaller parts and use what they already know to solve.

Think It Out!

Read each word problem. Then solve the number sentence written for that problem using any strategy you choose. Write the answer as a complete sentence. The first one has been done for you.

Read It!	Solve It!	Write It!
Batu ate 6 apples. Zola ate 2 fewer apples than Batu and 3 fewer apples than Chow. How many apples did Chow eat?	$6 - 2 = ? - 3$ $4 = ? - 3$ $4 = 7 - 3$	Chow ate 7 apples.
Tali had 7 pink shells. Kazoo had 7 more shells than Tali, but only 4 more shells than Gabby. How many shells did Gabby have?	$7 + 7 = ? + 4$	
Tali read 3 books. Zola read 6 more books than Tali but only 2 more than Rosie. How many books did Rosie read?	$3 + 6 = ? + 2$	

These problems help your child practice logic, a foundational critical thinking skill that requires your child to analyze the facts to solve a problem.

Skip Counting Back on a Hundred Chart

Counting Back Larger Numbers on a Hundred Chart

When you are subtracting an amount greater than 10, you will need to skip count back by a number that is not one. On a hundred chart, try skip counting back by tens.

Look at the problem 96 - 60 = ?

Break 60 into 6 tens. When you count back by tens, each number is ten less. Start at **96**. Skip count back 6 tens: **86**, **76**, **66**, **56**, **46**, **36**.

96 - 60 = **36**

1	2	3	4	5	6	7	8	9	10
11	12	13	14	15	16	17	18	19	20
21	22	23	24	25	26	27	28	29	30
31	32	33	34	35	36	37	38	39	40
41	42	43	44	45	46	47	48	49	50
51	52	53	54	55	56	57	58	59	60
61	62	63	64	65	66	67	68	69	70
71	72	73	74	75	76	77	78	79	80
81	82	83	84	85	86	87	88	89	90
91	92	93	94	95	96	97	98	99	100

- 10
- 10
- 10
- 10
- 10
- 10

When the number you are subtracting has both tens and ones, you can break up the number to skip count by tens **and** count back by ones to find the answer.

Look at the problem 73 - 42 = ?

Break the number you are subtracting into tens and ones: 42 is 4 tens and 2 ones.

First, skip count back by tens. Start at **73**. Skip count back 4 tens: **63**, **53**, **43**, **33**.

Next, count back the number of ones, starting at **33**: **32**, **31**.

73 - 42 = **31**

21	22	23	24	25	26	27	28	29	30
31	32	33	34	35	36	37	38	39	40
41	42	43	44	45	46	47	48	49	50
51	52	53	54	55	56	57	58	59	60
61	62	63	64	65	66	67	68	69	70
71	72	73	74	75	76	77	78	79	80

Count Back by Tens and Ones

Look at each problem. Use the hundred chart to skip count. Count back by tens or tens and ones to find the difference. Write the difference on the line provided. The first one has been done for you.

41	42	43	44	45	46	47	48	49	50
51	52	53	54	55	56	57	58	59	60
61	62	63	64	65	66	67	68	69	70
71	72	73	74	75	76	77	78	79	80
81	82	83	84	85	86	87	88	89	90
91	92	93	94	95	96	97	98	99	100

98 - 42 = 56

1	2	3	4	5	6	7	8	9	10
11	12	13	14	15	16	17	18	19	20
21	22	23	24	25	26	27	28	29	30
31	32	33	34	35	36	37	38	39	40
41	42	43	44	45	46	47	48	49	50
51	52	53	54	55	56	57	58	59	60

42 - 30 = ___

41	42	43	44	45	46	47	48	49	50
51	52	53	54	55	56	57	58	59	60
61	62	63	64	65	66	67	68	69	70
71	72	73	74	75	76	77	78	79	80
81	82	83	84	85	86	87	88	89	90
91	92	93	94	95	96	97	98	99	100

90 - 40 = ___

21	22	23	24	25	26	27	28	29	30
31	32	33	34	35	36	37	38	39	40
41	42	43	44	45	46	47	48	49	50
51	52	53	54	55	56	57	58	59	60
61	62	63	64	65	66	67	68	69	70
71	72	73	74	75	76	77	78	79	80

68 - 27 = ___

11	12	13	14	15	16	17	18	19	20
21	22	23	24	25	26	27	28	29	30
31	32	33	34	35	36	37	38	39	40
41	42	43	44	45	46	47	48	49	50
51	52	53	54	55	56	57	58	59	60
61	62	63	64	65	66	67	68	69	70

69 - 57 = ___

1	2	3	4	5	6	7	8	9	10
11	12	13	14	15	16	17	18	19	20
21	22	23	24	25	26	27	28	29	30
31	32	33	34	35	36	37	38	39	40
41	42	43	44	45	46	47	48	49	50
51	52	53	54	55	56	57	58	59	60

46 - 24 = ___

Hundred Chart Picture

Roar for More!

- Solve each subtraction problem by counting back on the hundred chart.
- Write the answer to the problem.
- Then, color the first number, the numbers used to skip count back by tens and/or ones, and the difference according to the color key to finish the picture on the opposite page.

Coloring Example: 67 - 53 = 14

1	2	3	4	5	6	7	8	9	10
11	12	13	**14**	**15**	**16**	**17**	18	19	20
21	22	23	24	25	26	**27**	28	29	30
31	32	33	34	35	36	**37**	38	39	40
41	42	43	44	45	46	**47**	48	49	50
51	52	53	54	55	56	**57**	58	59	60
61	62	63	64	65	66	**67**	68	69	70
71	72	73	74	75	76	77	78	79	80
81	82	83	84	85	86	87	88	89	90
91	92	93	94	95	96	97	98	99	100

Coloring Key

Orange

67 - 53 = ___
64 - 41 = ___
53 - 20 = ___
58 - 30 = ___
66 - 1 = ___
100 - 2 = ___
90 - 20 = ___

Yellow

55 - 30 = ___
56 - 30 = ___
97 - 3 = ___
86 - 11 = ___

Color Yellow

4, 7, 85

Green

92 - 91 = ___
91 - 80 = ___
60 - 52 = ___
89 - 71 = ___
84 - 11 = ___
88 - 11 = ___

Color any remaining squares green.

1	2	3	4	5	6	7	8	9	10
11	12	13	14	15	16	17	18	19	20
21	22	23	24	25	26	27	28	29	30
31	32	33	34	35	36	37	38	39	40
41	42	43	44	45	46	47	48	49	50
51	52	53	54	55	56	57	58	59	60
61	62	63	64	65	66	67	68	69	70
71	72	73	74	75	76	77	78	79	80
81	82	83	84	85	86	87	88	89	90
91	92	93	94	95	96	97	98	99	100

Mystery Picture

Color the picture according to the directions on the previous page.

Materials needed: crayons or colored pencils in orange, yellow, and green

The Rainforest

Your child is ready to read *The Rainforest*. This creative story combines both fiction and nonfiction elements, as well as a visual representation of counting down by tens on a vertical number line.

During reading:

Talk about the difference between fiction and nonfiction. Point out that the photographs show real plants and animals, while the illustrations show an imaginary creature. Then have your child look at the photos to see if they can spot the hidden "fozzle" on each set of pages.

Point out the number line as you go through the story. Ask your child what number is being used to skip count down through the rainforest. Can they predict what number will come next in the story?

After reading:

Ask your child what imaginary creature they would like to invent. They can draw it with colored pencils, create it with clay, or even design it with a computer drawing program! What sounds would their creature make? What name would they give it?

Celebrate your child's progress by placing a star sticker in the workbook.

To the Finish Line

The pals are donating books to the library. How many books will each pal have left? Subtract on the number line to see. First, circle the numbers in the problem. Then draw and count the jumps between the numbers. Write the answer. Then circle the pal who donates the most books.

Tali had 52 books. After giving some to the library, she had 46 left. How many books did she donate?

43 44 45 46 47 48 49 50 51 52 53 54

52 - 46 = ___

Zola had 35 books. After giving some to the library, she had 29 left. How many books did she donate?

27 28 29 30 31 32 33 34 35 36 37 38

35 - 29 = ___

Kazoo had 76 books. After giving some to the library, he had 67 left. How many books did he donate?

66 67 68 69 70 71 72 73 74 75 76 77

76 - 67 = ___

Your child has completed Step 5!
Celebrate by placing a star sticker in the workbook.

Launch

Modeling Single-digit Subtraction with Regrouping

When subtracting a single-digit number from a double-digit number, you start with the ones. But if there are not enough ones, you must *regroup* and change one group of ten into 10 ones.

Place value rods and blocks can model subtraction problems that include regrouping. Find the difference: 23 - 4 = ?

23 is 2 tens rods and 3 ones blocks. **4** is 4 ones blocks. Since 4 can't be subtracted from 3, regrouping is necessary.	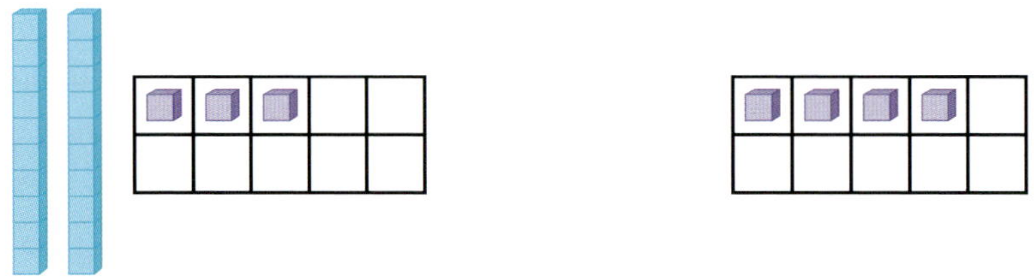 **23** - **4** = ?
Here, regrouping means changing a tens rod into a group of 10 ones blocks. The **23** is regrouped into 1 ten and 13 ones blocks (**10 + 13**).	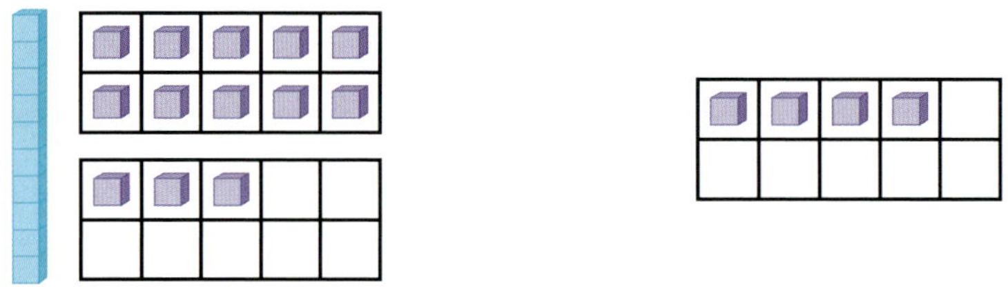 **10 + 13** - **4** = ?
After regrouping, you can take away 4 ones. That leaves 1 tens rod and 9 ones blocks. So, 23 - 4 = **19**	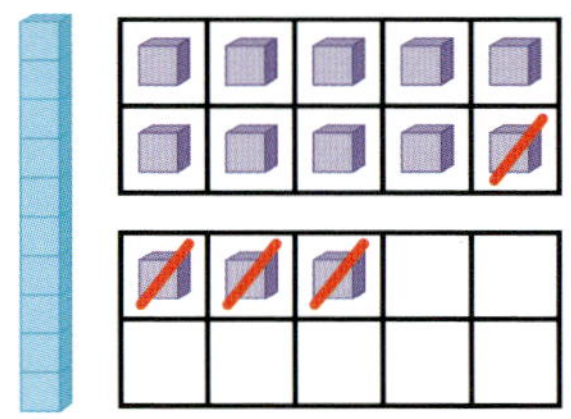 **23 - 4 = 19**

Stack coins, trading cards, or any small object in sets of 10. Ask your child to subtract a smaller number from the sets. Whenever they must break up a set of 10, they are regrouping.

Once Around the Block

Solve each place value block subtraction sentence by regrouping. Cross out a tens rod and draw it as 10 ones. Then cross out the number of blocks being taken away. Write the subtraction sentence. The first one has been done for you.

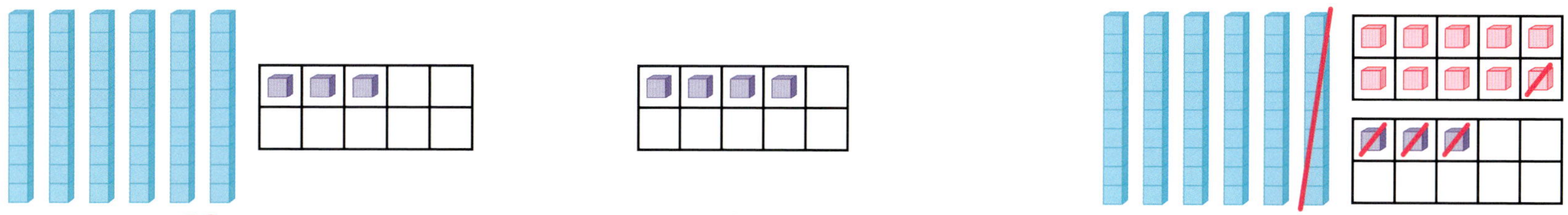

63 - 4 = 59

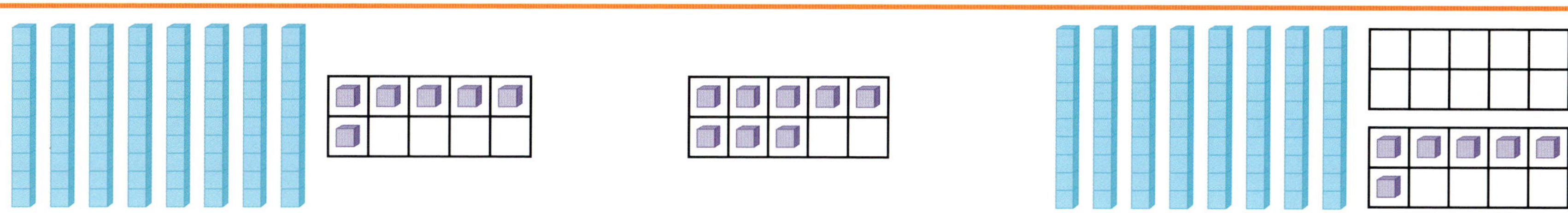

___ - ___ = ___

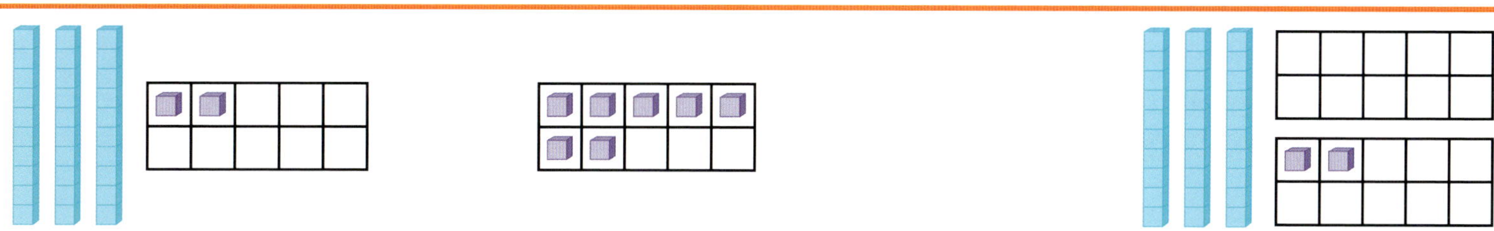

___ - ___ = ___

Encourage your child to go back and check their answers on a hundred chart, counting back by ones, which they learned on page 12.

Launch

Modeling Double-digit Subtraction with Regrouping

Place value rods and blocks can also model double-digit subtraction problems that include regrouping.

Find the difference: 42 - 24 = ?

	Place Value Blocks Number Sentence
42 is 4 tens rods and 2 ones blocks. **24** is 2 tens rods and 4 ones blocks. Start with the ones. Since 4 can't be subtracted from 2, regrouping is necessary.	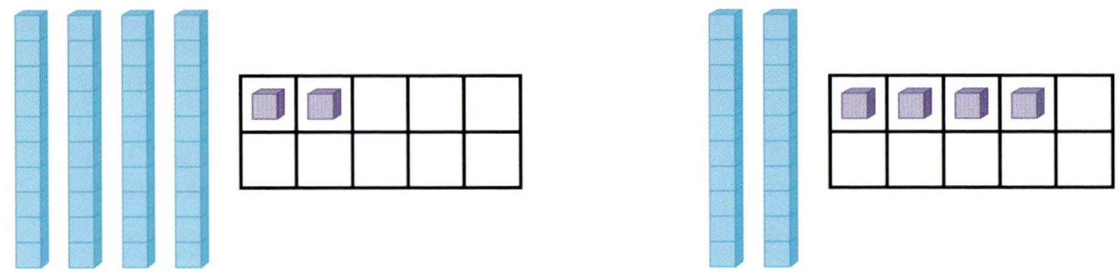 **42** - **24** = ?
Here, regrouping means changing a tens rod into a group of 10 ones blocks. The **42** is regrouped into 3 tens rods and 12 ones blocks (**30 + 12**).	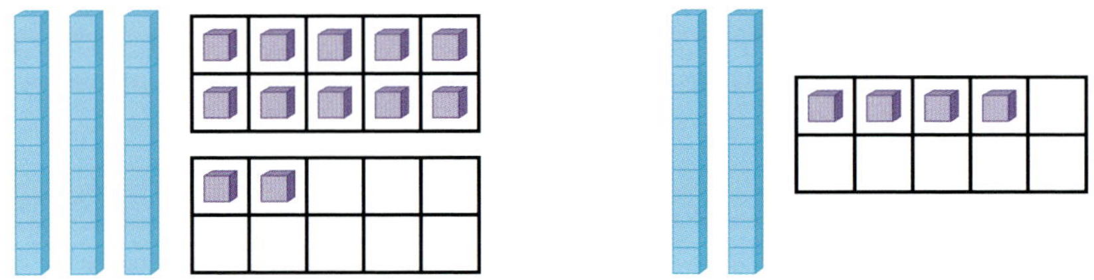 **30 + 12** - **24** = ?
After regrouping, you can take away 4 ones. That leaves 3 tens rods and 8 ones blocks. Next, take away the 2 tens from the remaining 3 tens. That leaves 1 ten and 8 ones. So, 42 - 24 = **18**	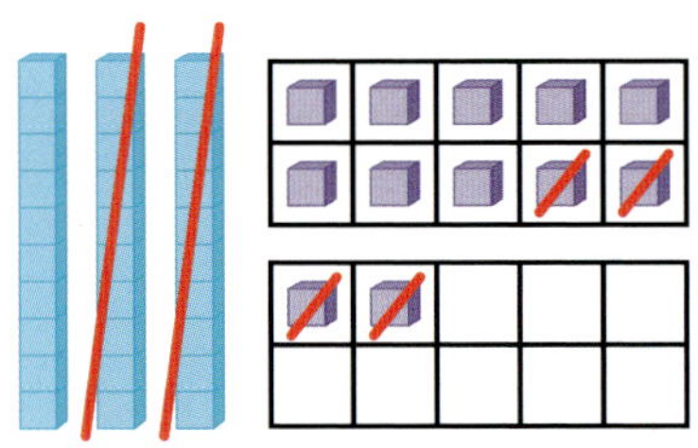 **42 - 24 = 18**

Around the Block Again

Solve each place value block subtraction sentence by regrouping. Cross out a tens rod and draw it as 10 ones. Then cross out the number of blocks and rods being taken away. Write the subtraction sentence. The first one has been done for you.

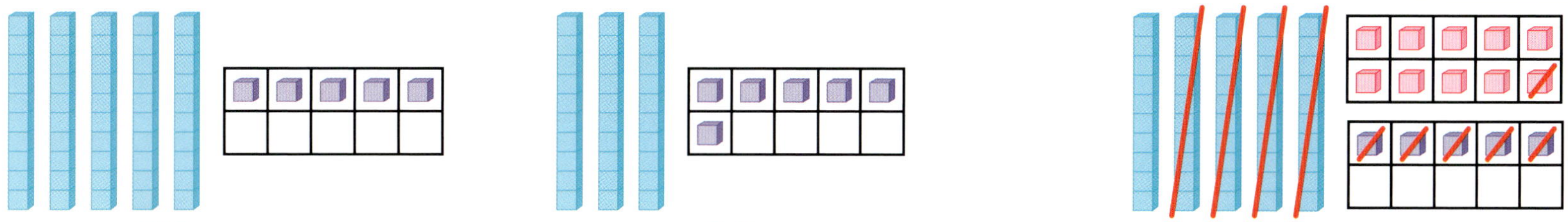

55 - 36 = 19

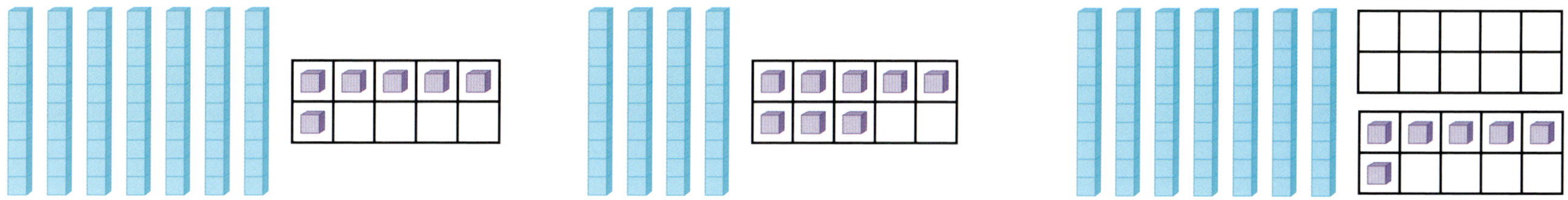

___ - ___ = ___

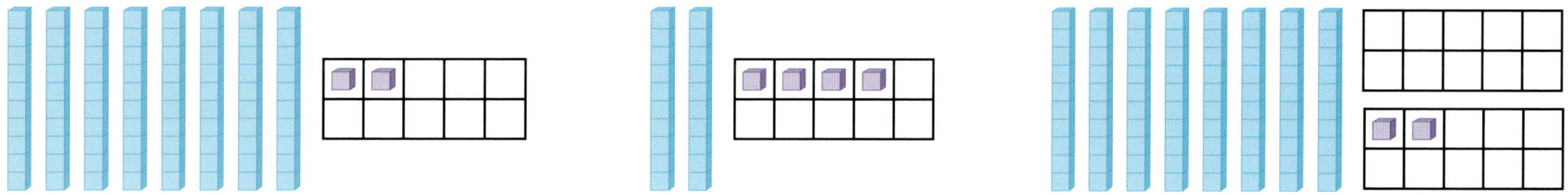

___ - ___ = ___

Encourage your child to go back and check their answers on a hundred chart, counting back by tens and ones, which they learned on page 16.

Vertical Subtraction with Regrouping

Vertical subtraction lines up the numbers in columns according to their place values. This helps you to subtract the numbers in each column separately to find the difference.

Find the difference: 56 - 38 = ?

First, write the two numbers as a vertical subtraction sentence.

Tens	Ones
5	6
- 3	8

Start subtracting from the ones column. Notice that the bottom number, **8**, is larger than the top number, **6**. You cannot take 8 away from 6.

Regroup a set of ten as 10 ones. Draw a line through the **5** and write a **4** to show you have regrouped 1 ten.

Tens	Ones
4 ~~5~~	16
- 3	8

Next, add the 10 ones you are regrouping to the 6 ones. Since 10 + 6 = 16, write the number **1** next to the **6** to show that there are now 16 ones.

Now, subtract the ones:
16 - 8 = **8**

Next, subtract the tens:
4 - 3 = **1**

Tens	Ones
4 ~~5~~	16
- 3	8
1	8

1 ten and 8 ones is equal to 18.

56 - 38 = **18**

Regrouping in subtraction is sometimes called *borrowing*. You may wish to remind your child that in the App, we refer to this as "go next door and take ten more."

Take It Away!

Solve each subtraction problem. Then match the letter to the blanks below to find the answer to the riddle.

71 − 52 = ___ = A	64 − 49 = ___ = B	82 − 57 = ___ = E	34 − 17 = ___ = I
31 − 15 = ___ = K	55 − 26 = ___ = N	95 − 68 = ___ = R	41 − 23 = ___ = V

What kind of bank has no money?

___   ___ ___ ___ ___ ___   ___ ___ ___ ___

19   27 17 18 25 27   15 19 29 16

What's the Word?

Follow the same steps you learned earlier to solve word problems—just remember to regroup when you're subtracting!

Read the problem and search for key words and information.
Set up a horizontal math number sentence, then switch it to be vertical. When regrouping, remember to cross out a group of 10 in the tens column and add it to the group of ones.
Answer the question using a complete sentence. Remember to hyphenate any number between 20 and 99.

Kazoo had **forty-six** blocks. He gave **seventeen** of them to Tali. How many blocks did he have **left**?

46 - 17 = ?

$$\begin{array}{r} {}^{3}\not{4}{}^{1}6 \\ -\ 17 \\ \hline 29 \end{array}$$

Kazoo had **twenty-nine** blocks left.

Reverse!

Write a word problem for the following equation: **65 - 28 = ?** Then solve it!

$$\begin{array}{r} 65 \\ -\ 28 \\ \hline \end{array}$$

____

____

Review Lesson 1 of Step 3 if your child needs help remembering key words.

A Careful Reading

Read each word problem below. Then:

- Circle the numbers you need to work with in the problem.
- Underline the words that tell what operation to perform.
- Write each number sentence horizontally, and then vertically.
- Solve the problem and write the answer to the question as a complete sentence.

Story	Horizontal	Vertical	Answer
Batu had seventy-four tickets at the fair. He gave thirty-six to Chow so that Chow could get a prize. How many tickets did Batu have left?			
Kazoo delivered eighty-two packages, and forty-seven of them were for Gabby. How many of the packages were not for Gabby?			
The lemon tree had ninety-three lemons growing on it. Tali picked fifty-four of them and put them in a basket. How many lemons were left on the tree?			

Place Value Blocks Subtraction Number Sentences

Place value blocks can help you see subtraction in action. Remember that place value is the value of each digit in a number.

Look at the subtraction sentence 320 - 217 = ?

Don't Forget!

Flat = 100 Rod = 10 Block = 1

320 - 217 = ?

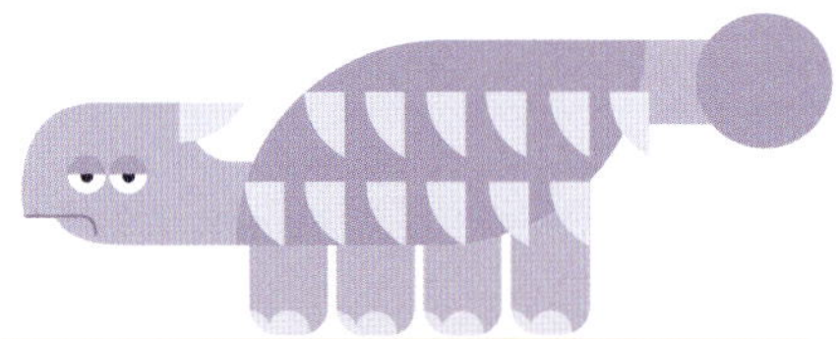

To regroup the number 320, cross out 1 tens rod and draw 10 ones blocks.

Start with the ones. Cross out the numbers you are subtracting. Continue to take away numbers by crossing out the tens, then the hundreds.

320 - 217 = **103**

Remember, the number zero acts as a placeholder if there are no tens or ones. If there are no hundreds, there is no need to write the 0.

Hold My Place

Two- and three-digit numbers are represented by flats, rods, and blocks.

- In the first two rows, write the value of the flats, rods, and/or blocks, including a 0 if needed, to write the vertical subtraction sentence.
- In the last row, cross out a tens rod and draw it as 10 ones if needed. Then cross out the rods and blocks to take them away.
- Write the answer. The first one has been done for you.

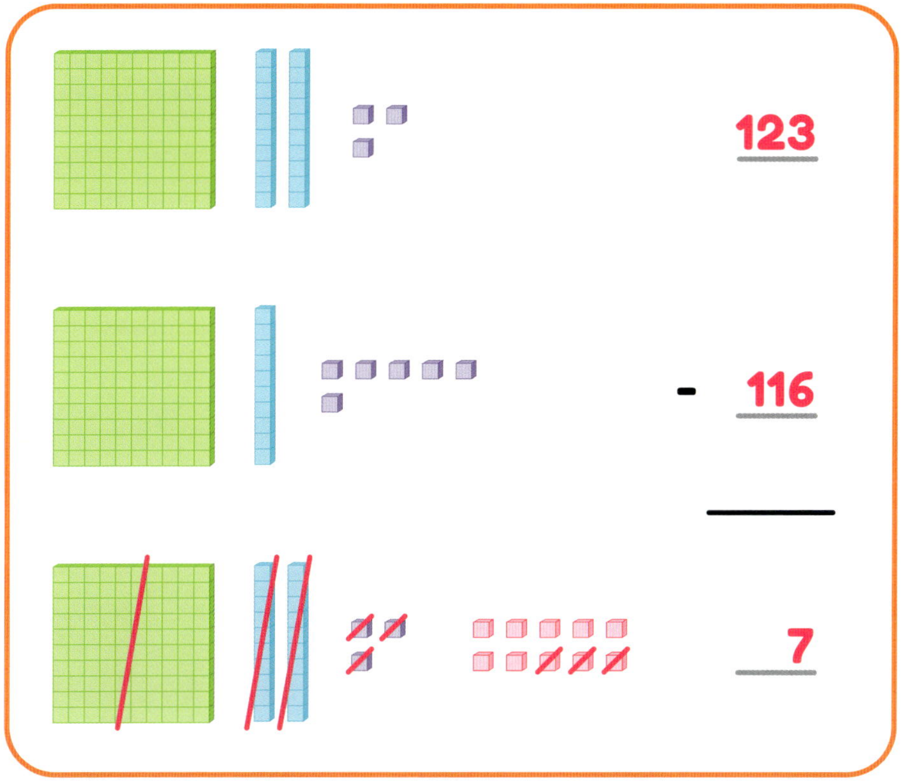

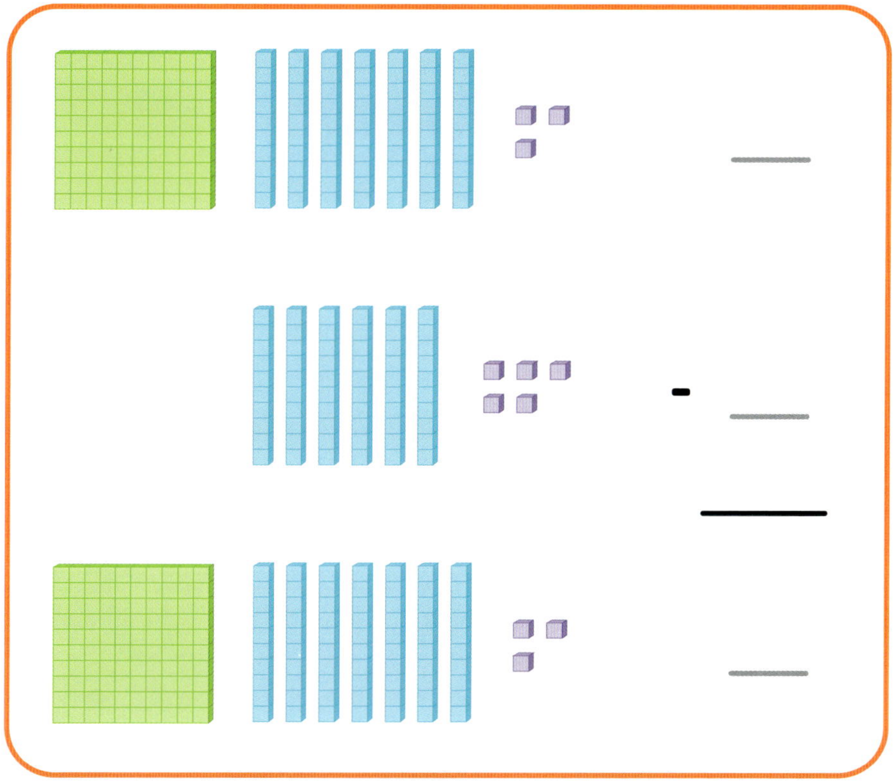

Find the Flowers

Solve the subtraction problems. Color the differences that are even numbers **green** to help the bee find the flowers. **Hint:** you may need to move up, down, or across. The first one has been done for you.

Start

2 ~~3~~	15
- 1	7
1	8

7	5
- 1	6

4	5
- 1	8

6	1
- 3	6

8	3
- 4	6

5	7
- 1	9

2	2
- 1	7

8	1
- 3	3

6	5
- 3	9

4	2
- 1	4

4	6
- 1	8

9	4
- 7	6

5	3
- 2	5

7	2
- 3	7

2	1
-	5

Finish

Back to the Hive

Solve the subtraction problems. Color the differences that are odd numbers **yellow** to help the bee get back to its hive. **Hint:** you may need to move up, down, or across.

Start →

66 - 47 ——	72 - 39 ——	93 - 87 ——	56 - 48 ——	85 - 47 ——
77 - 39 ——	32 - 17 ——	86 - 28 ——	57 - 29 ——	62 - 24 ——
94 - 76 ——	36 - 29 ——	88 - 49 ——	34 - 15 ——	94 - 57 ——

Finish

The Grand Meerkat Hotel

Your child is ready to read *The Grand Meerkat Hotel*, a whimsical story that features double-digit subtractio problems with regrouping. To reinforce this concept, color-coded numbers help your child easily follow the subtraction equations.

During reading:

Occasionally cover up the answer to one of the subtraction problems in the book. Ask your child to explain the steps that are required to solve it.

Discuss the idea of regrouping. Make sure your child understands that the digits represent the number in that place, and that each place can only hold a single digit.

After reading:

Organize a family reunion—even if it only includes those members who live with you. Ask your child to design an invitation to promote this event. What fun activities can you do all together?

Like Kazoo in the story, has your child ever felt like a “terrible helper” to someone? What did they do? How did they resolve the problem?

What kind of help does your child like to give? Are they an “expert helper” in any household or school chores?

Celebrate your child’s progress by placing a star sticker in the workbook.

Help Kazoo Win a Prize!

Kazoo is playing games at the arcade. Help him keep track of the tickets he needs to win. After you solve each subtraction problem, write the difference, then write it again as the number to take away from in the next game. If you reach 0, Kazoo will win a prize! The first one has been done for you.

Game 1

Kazoo needs to earn 95 tickets. He wins 29 tickets.
How many tickets does he still need?

Game 2

Kazoo wins 28 tickets.
How many tickets does he still need?

Game 3

Kazoo wins 19 tickets.
How many tickets does he still need?

Game 4

Kazoo wins 19 tickets. Did he win a prize?

Game 1

8 ~~9~~	1 5
− 2	9
6	6

Game 2

6	6
−	

Game 3

−	

Game 4

−	

Hooked on Math

It's Time to Celebrate!

____

has completed Elementary Math Steps 5 & 6 and is now ready to move on to the Elementary Math Steps 7 & 8 Workbook!

SKILLS MASTERED

- Subtracting on a number line
- Counting back on a hundred chart
- Skip counting back on a hundred chart
- Subtraction with regrouping
- Subtraction word problems with regrouping
- Place value blocks subtraction

5 6 7 8

Don't forget to share your success and tag us on social media so we can celebrate with you!

Credits

Design Team
Julie Linden Frazier
Lindsey Frazier

Editorial Team
Donna Giachetti
Cynthia Stierle
Austin Feld

Illustrations
Dubit Limited

Educational Advisor
Lesley Freed, Curriculum Specialist

Published by Sandviks HOP, Inc.

Printed in China

Hooked on Math

Elementary Math

18040 (11/22)

hookedonmath.com

Hooked on Math

Elementary Math

Steps 3 & 4

Workbook

Tips for Success

- Elementary Math focuses on understanding the importance of place value while working with numbers up to 1,000. It also introduces the key concept of regrouping and builds on the addition and subtraction skills introduced in Math Fundamentals.
- Work at a pace that's right for you and your child.
- Fit your session length to your child's attention span. It's okay to work with the App one day and this workbook the next.
- Before beginning, flip through the workbook and note any materials you might need. You should plan to have pencils, a pair of dice, and crayons ready.
- You'll see parent tips throughout this workbook. These tips are designed to answer questions you or your child might have about the learning.
- Repetition and review are essential to basic math fluency. Each workbook step builds on the ones before, so make sure your child has mastered the current lesson before moving forward.
- As you go through the workbook, encourage your child to explore the connection between numbers and the real world.
- Have fun with it!

Hooked on Math

Elementary Math

Steps 3 & 4

Workbook

This workbook and the included storybooks introduce elementary math skills.

Step 3: Strategies to 100
- Adding tens and ones word problems
- Skip counting on a hundred chart
- Making numbers using place value blocks
- Addition strategies review

Step 4: Addition within 100
- Regrouping to add
- Word problems with regrouping
- Part-part-whole review

Hooked & Company

Contents

Lesson Overview

Welcome! This workbook provides supplemental activities that will solidify and expand the learning your child is enjoying with the Hooked on Math App. Follow the easy, step-by-step instructions to guide your child's exploration of numbers and operations. Every lesson helps your child build a strong foundation in elementary math skills while having fun!

1. Launch

Your child will be introduced to math concepts and skills through interactive lessons in the App on your mobile device or Windows or Mac computer.

Download the App at: iOS

hookedonmath.com/get-the-app

2. Discover

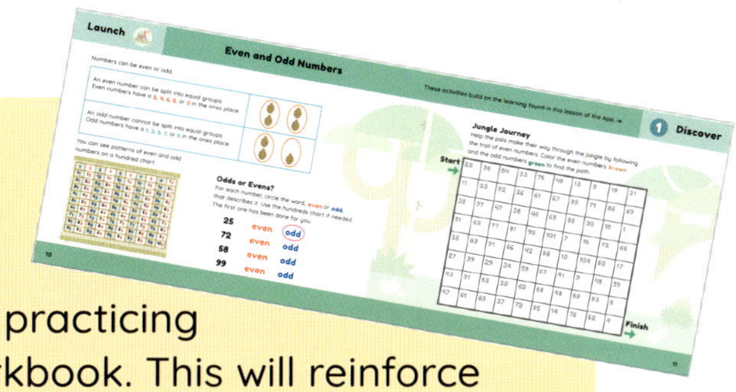

Next, your child will have fun exploring math concepts and practicing new skills in the workbook. This will reinforce the App lessons and provide a tangible application of the skills learned.

3. Play

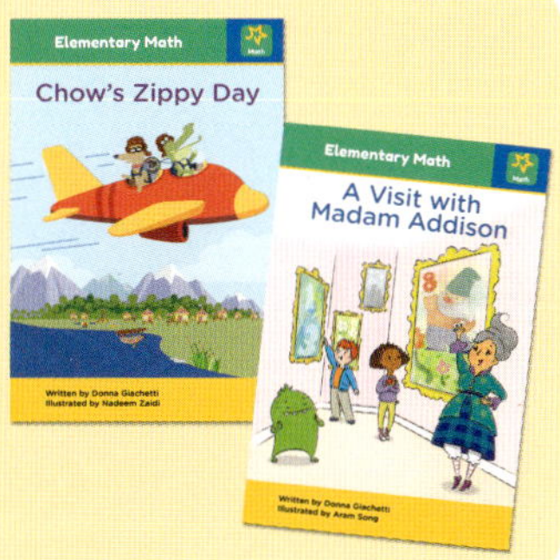

In addition to the workbook activities, your child will enjoy two math-based storybooks that will strengthen their learning and add to the fun!

4. Celebrate

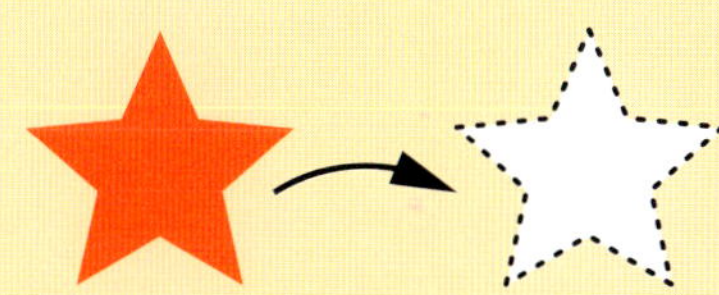

Celebrate your child's progress by placing a sticker in the workbook as learning milestones are reached. This also provides a helpful marker to remember where to begin next time.

Special notes are included in this space to provide additional context related to what is being taught.

Review

Even and Odd Numbers

Mystery Number

Help the Great Chowdini guess the mystery number.

- Read each clue.
- Cross out numbers on the hundred chart that are **not** the mystery number.
- Circle the mystery number and write it on the line below.

Clues

1. It is *not* an odd number.
2. It is greater than 36.
3. It is less than 82.
4. There is a 5 in the tens place.
5. There is an 8 in the ones place.

The mystery number is ______

1	2	3	4	5	6	7	8	9	10
11	12	13	14	15	16	17	18	19	20
21	22	23	24	25	26	27	28	29	30
31	32	33	34	35	36	37	38	39	40
41	42	43	44	45	46	47	48	49	50
51	52	53	54	55	56	57	58	59	60
61	62	63	64	65	66	67	68	69	70
71	72	73	74	75	76	77	78	79	80
81	82	83	84	85	86	87	88	89	90
91	92	93	94	95	96	97	98	99	100

Coconut Clusters

Split the coconuts into equal groups of your choosing. Then write how many groups and the number in each group. **Hint:** there is more than one way to split each group.

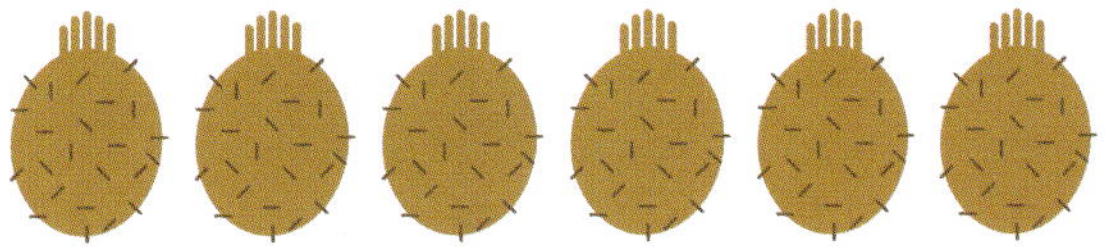

How many groups? _____

How many in each group? _____

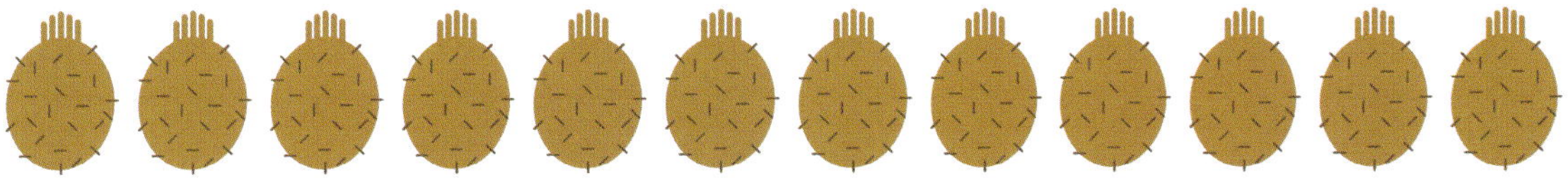

How many groups? _____

How many in each group? _____

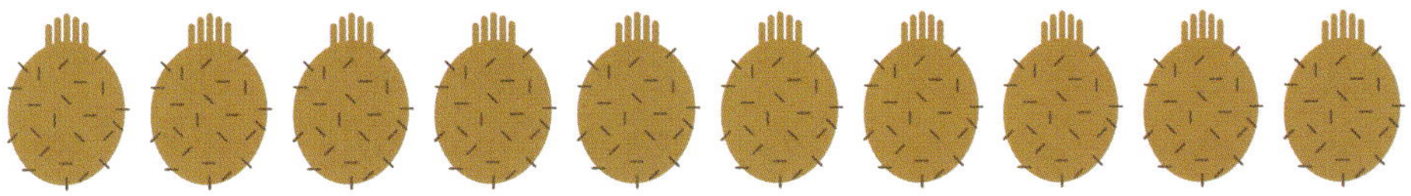

How many groups? _____

How many in each group? _____

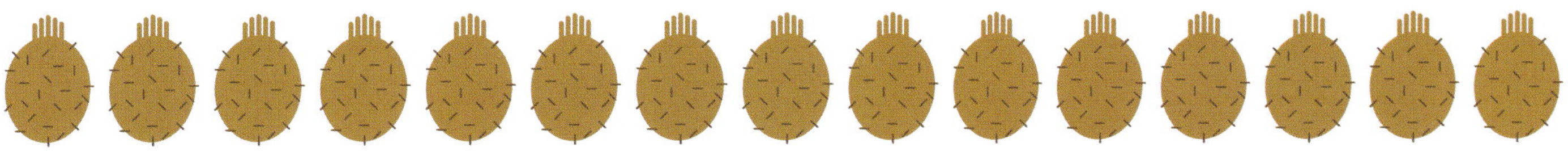

How many groups? _____

How many in each group? _____

Donut Count

How many donuts are on each tray? Write two different repeated addition sentences for each array and fin the sum.

Number Cloud Find

You know that numbers can be added in any order and still have the same sum. Each sentence in the chart has been written three different ways in the number cloud. When you find a sentence in the number cloud that has the same sum as one in the chart, cross it out and make a check mark on the chart. The first one has been done for you.

17 + 6 + 4 = 27	✓	✓	✓
8 + 9 + 12 = 29			
5 + 6 + 15 = 26			
14 + 3 + 7 = 24			
1 + 9 + 5 = 15			

~~6 + 4 + 17 = 27~~ 7 + 3 + 14 = 24

12 + 9 + 8 = 29

9 + 8 + 12 = 29

~~4 + 6 + 17 = 27~~

8 + 12 + 9 = 29

9 + 1 + 5 = 15

3 + 7 + 14 = 24

15 + 5 + 6 = 26

5 + 15 + 6 = 26

5 + 9 + 1 = 15

14 + 7 + 3 = 24

~~4 + 17 + 6 = 27~~

1 + 5 + 9 = 15 6 + 5 + 15 = 26

A Wheel Puzzle

Can you think of ten ways to make three numbers add up to 25? You can use any combination of three numbers **and it's okay to repeat numbers**.

- If you write a sentence that uses the Make 10 strategy, put a check mark next to the first number.
- If you use the doubles strategy, put two check marks next to the first number.
- The first one has been done for you.

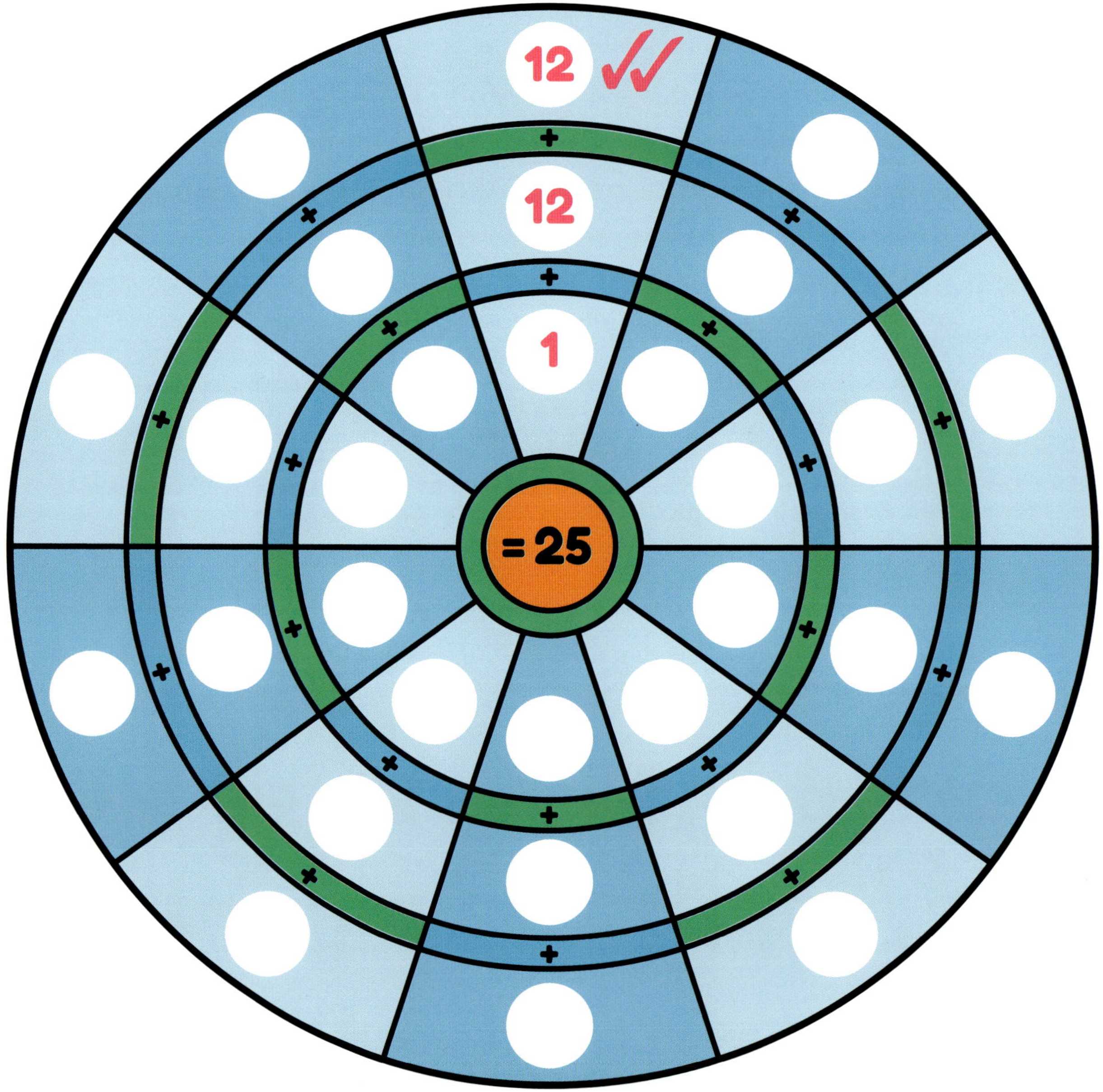

Blow Its Top!

Find and circle the lava stream that completes each number bond. Then write the subtraction sentence. The first one has been done for you.

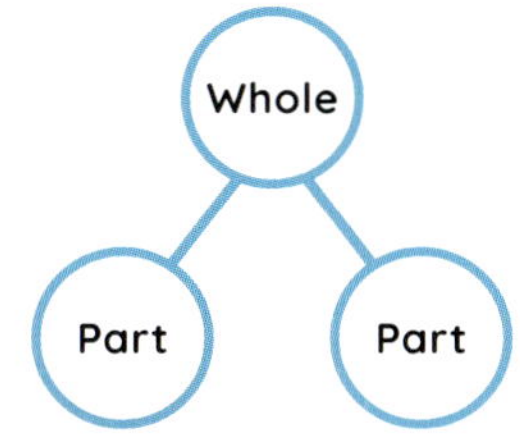

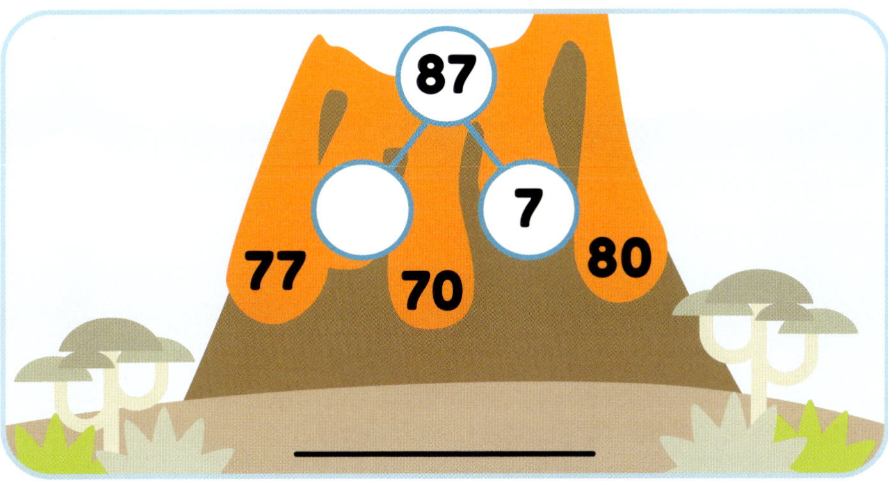

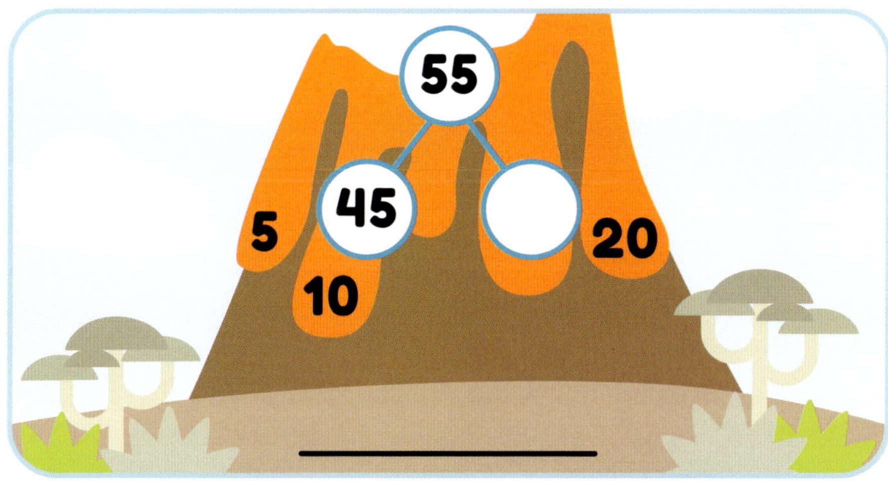

Adding Tens and Ones Word Problems

Word Problem Detective

To solve a word problem, read the story and search for the information you need to find the answer.

1. Look for information that includes the numbers to work with.
2. Look for words or phrases that tell you whether you need to do addition or subtraction.
3. Set up a horizontal math number sentence.
4. Switch it to a vertical equation and solve it.
5. Answer the word problem question in a complete sentence.

There are **thirty-six** coconuts and **thirteen** oranges in the basket. How many fruits **in all**?

$36 + 13 = ?$

$$\begin{array}{r} 36 \\ +\ 13 \\ \hline 49 \end{array}$$

There are 49 fruits in the basket.

Give Me a Sign

Read the words or phrases. Circle the + or – sign to show which operation should be performed.

All together	+	–
Fewer than	+	–
Take away	+	–
In all	+	–

\+
For addition, search for keywords and phrases like *all together, join,* or *in all.*

–
For subtraction, search for keywords and phrases like *take away, left over* or words that end in the letters *-er,* like *fewer.*

In a Few Words

Read the story in each word problem below.

- Circle the numbers you need to work with in the story.
- Underline the words that tell what operation to perform.
- Write each number sentence horizontally, and then vertically.
- Solve the problem and write the answer to the question.

  The first one has been done for you.

Story	Horizontal	Vertical	Answer
There are thirty-three pals in line for the show. Twelve more pals join the line. How many pals are in the line all together?	33 + 12 = ?	33 + 12 45	There are 45 pals in the line.
The parking lot can hold forty-six cars. There are twenty-two cars parked in the lot now. How many empty spots are left over?			
The train traveled seventy-five miles. It stopped at the station, then traveled twenty-four more miles. How many miles did it travel in all?			

Launch

Skip Counting Review

In math, skip counting means to count forward by a number that is not 1. To skip count, add the same number each time to the number that comes before it. You can skip count from any number—you don't have to start at 0.

To skip count by 2s, add 2 to each number to get to the next number. For example, **10**, **12**, **14**, **16**, **18**.

To skip count by 5s, add 5 to each number to get to the next number. For example, **25**, **30**, **35**, **40**, **45**, **50**.

To skip count by 10s, add 10 to each number to get to the next number. For example, **47**, **57**, **67**, **77**, **87**.

Out of Step

Three of the patterns in each row show skip counting by the same number. One pattern is different. Circle the pattern in each row that **doesn't** belong.

18, 20, 22, 24, 26	33, 35, 37, 39, 41	33, 36, 39, 42, 45	82, 84, 86, 88, 90
35, 40, 45, 50, 55	62, 72, 82, 92, 102	75, 80, 85, 90, 95	13, 18, 23, 28, 33
10, 20, 30, 40, 50	23, 33, 43, 53, 63	61, 71, 81, 91, 101	105, 110, 115, 120, 125

Pretty Perplexing Patterns

Each color highlights a skip counting pattern of either 2s, 5s, or 10s on the hundred chart.

- Write the number pattern next to the color name.
- Circle whether the pattern is skip counting by 2s, 5s, or 10s.
- The first one has been done for you.

Color	Number Pattern	Skip Count
Yellow	12, 22, 32, 42	2s 5s (10s)
Pink		2s 5s 10s
Green		2s 5s 10s
Blue		2s 5s 10s
Red		2s 5s 10s
Aqua		2s 5s 10s
Purple		2s 5s 10s

1	2	3	4	5	6	7	8	9	10
11	12	13	14	15	16	17	18	19	20
21	22	23	24	25	26	27	28	29	30
31	32	33	34	35	36	37	38	39	40
41	42	43	44	45	46	47	48	49	50
51	52	53	54	55	56	57	58	59	60
61	62	63	64	65	66	67	68	69	70
71	72	73	74	75	76	77	78	79	80
81	82	83	84	85	86	87	88	89	90
91	92	93	94	95	96	97	98	99	100

Making Numbers Using Place Value Blocks

Remember that place value is the value of each digit in a number.

Don't Forget!

Flat = 100 Rod = 10 Block = 1

A Valuable Place

Look at the flats, rods, and blocks below. Write how many each one represents in the box below it. Then write what number they represent all together. The first one has been done for you.

	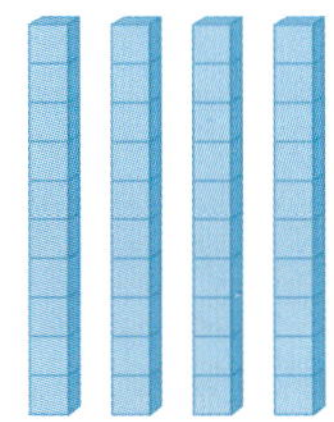	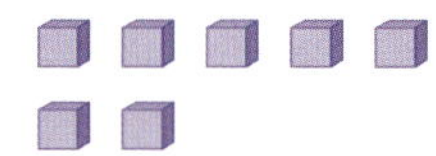
200	40	7
	247	

	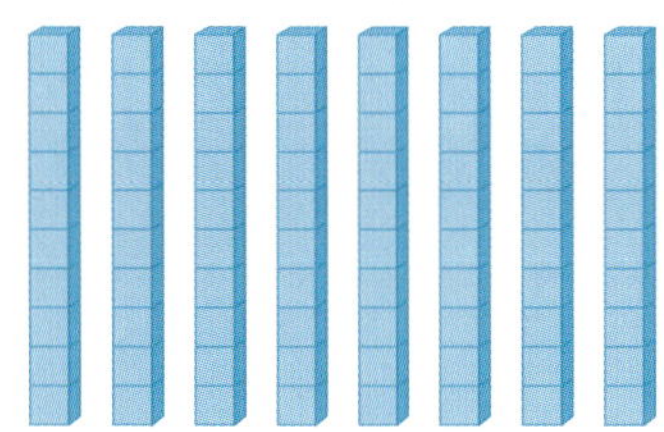	

What's My Value?

A three-digit number can be represented by flats, rods, and blocks. Write the entire number as a numeral in the Numeral column. Then write the value of the circled flats, rods, or blocks in the Digit Value column. The first one has been done for you.

	Numeral	Digit Value
	146	40
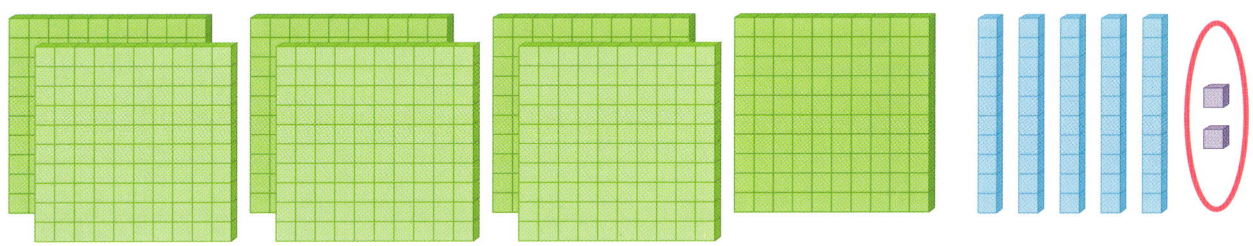		

Launch

Place Value Addition Strategy Review

You can use place value strategies to help you add.

Hundred Chart

You can use the down and across strategy to add on a hundred chart.

42 + 34 = ?

- Break one of the numbers you are adding into tens and ones: 34 is 3 tens and 4 ones.
- First, skip count down by tens. Start at **42**. Skip count down 3 tens: **52**, **62**, **72**.
- Next, count across the number of ones, starting at **72**: **73**, **74**, **75**, **76**.

42 + 34 = **76**

	1	2	3	4	5	6	7	8	9	10
	11	12	13	14	15	16	17	18	19	20
	21	22	23	24	25	26	27	28	29	30
	31	32	33	34	35	36	37	38	39	40
	41	42	43	44	45	46	47	48	49	50
+ 10	51	52	53	54	55	56	57	58	59	60
+ 10	61	62	63	64	65	66	67	68	69	70
+ 10	71	72	73	74	75	76	77	78	79	80
	81	82	83	84	85	86	87	88	89	90
	91	92	93	94	95	96	97	98	99	100
			+1	+1	+1	+1				

Breaking Up Isn't Hard to Do

Break numbers into tens and ones to add them.

65 + 24 = ?

- Break each number into tens and ones: **60** + **5** and **20** + **4**
- Add the tens.
- Add the ones.
- Add the sums.

60	+	20	=	80	Tens
5	+	4	=	9	Ones
				89	Sum

Decomposing and composing numbers simply means breaking numbers apart and putting them back together. This can help with the concept of regrouping.

On the Chart

Find the sum for each problem and write the answer. Use the down and across strategy on the hundred chart to help you.

37	41	75	23	52
+ 21	+ 16	+ 24	+ 66	+ 17

1	2	3	4	5	6	7	8	9	10
11	12	13	14	15	16	17	18	19	20
21	22	23	24	25	26	27	28	29	30
31	32	33	34	35	36	37	38	39	40
41	42	43	44	45	46	47	48	49	50
51	52	53	54	55	56	57	58	59	60
61	62	63	64	65	66	67	68	69	70
71	72	73	74	75	76	77	78	79	80
81	82	83	84	85	86	87	88	89	90
91	92	93	94	95	96	97	98	99	100

Stretch It Out

Look at each addition sentence. Expand each number into tens and ones to add across, then add down to find the sum. The first one has been done for you.

53 + 24 = ?

50	+	20	=	70
3	+	4	=	7
				77

32 + 44 = ?

	+		=	
	+		=	

21 + 68 = ?

	+		=	
	+		=	

What Is the Question?

To find the missing number in each addition problem, look at the section of the hundred chart. How many tens and ones are added to the starting number to reach the sum? Go down to count tens and across to count ones. Then write the missing number in the equation. The first one has been done for you.

15 + 42 = 57

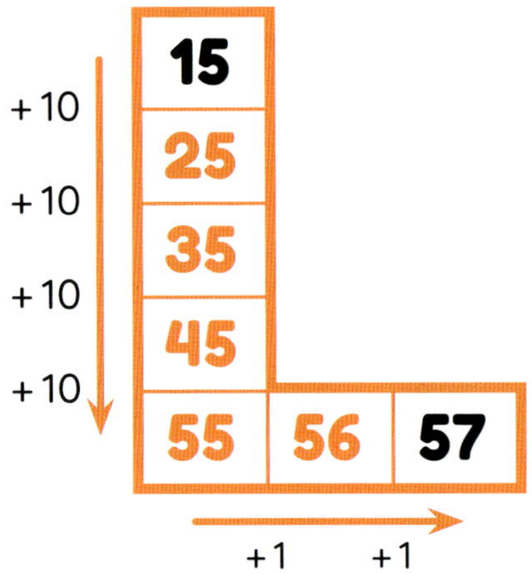

84 + ___ = 95

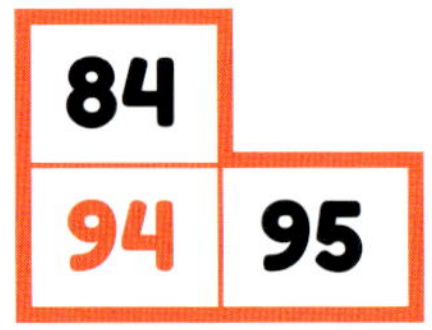

61 + ___ = 97

13 + ___ = 47

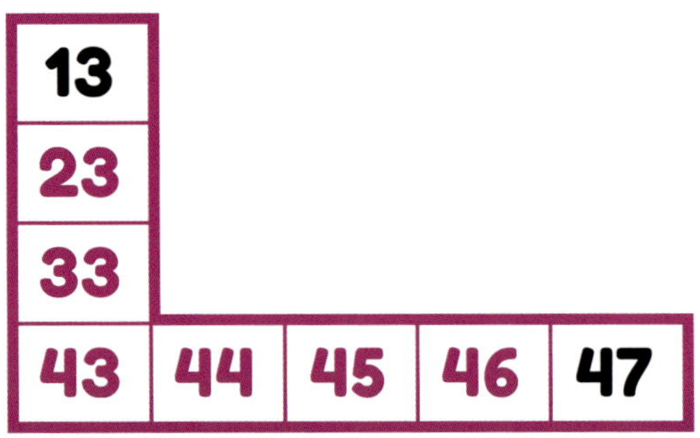

25 + ___ = 49

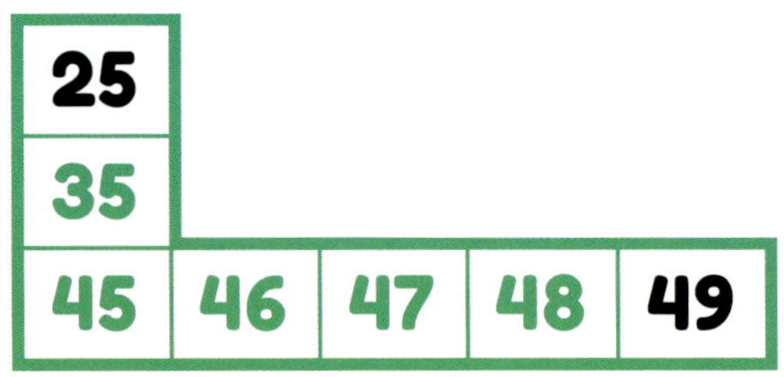

32 + ___ = 86

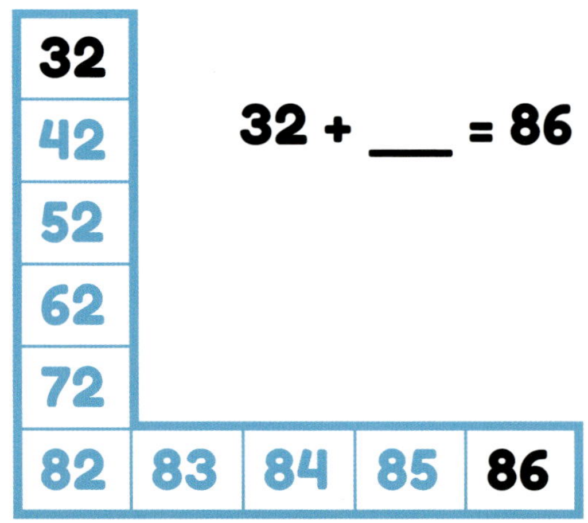

Check the List

For each number on the list, find the three boxes on the shelf that add up to that number. When you find a box with the same value as one of the digits on the list, color it the same color as the number on the list. The first one has been done for you.

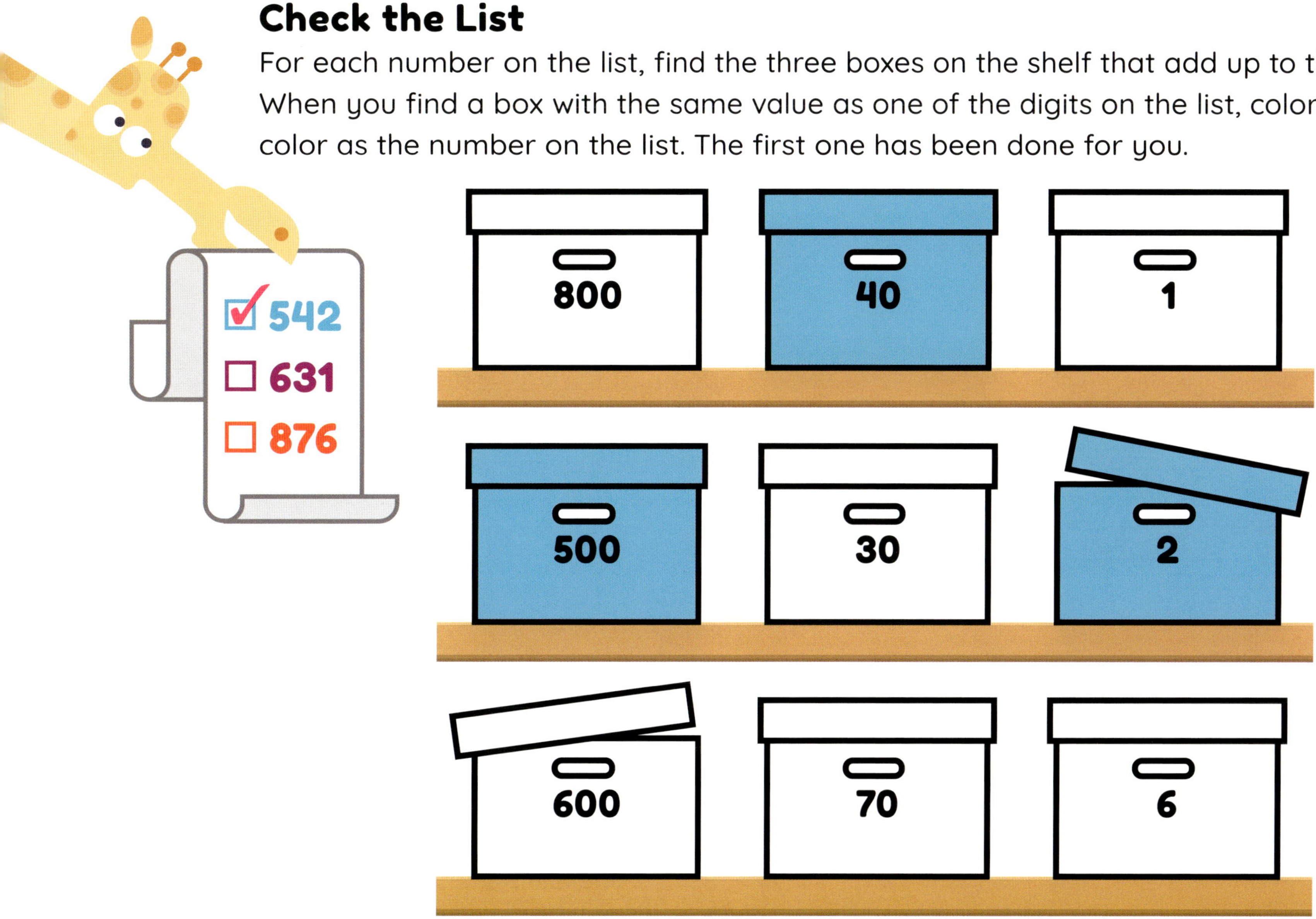

Materials needed: crayons or colored pencils in blue, purple, and red

Chow's Zippy Day

Your child is ready to read *Chow's Zippy Day*. This amusing story includes visual representations of one-digit, two-digit, and three-digit numbers. Throughout the story, the fictional speed measurement of "zips per hour" is represented with numbers in the place value chart as well as with flats, rods, and blocks.

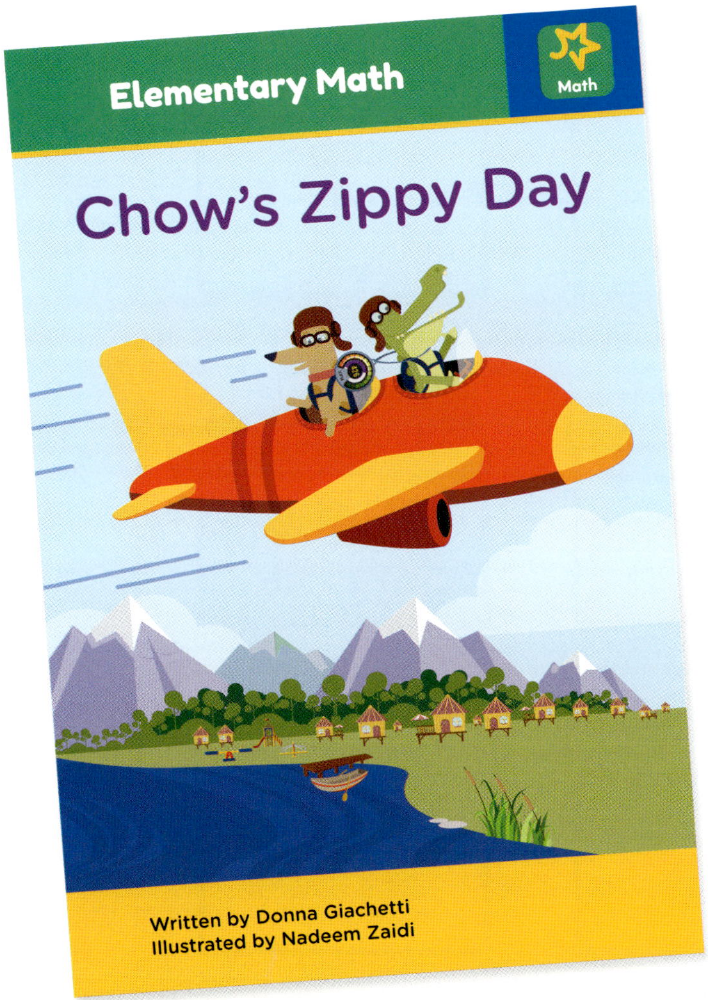

Understanding that where a digit is placed within a number relates to an amount is a key concept at this stage, especially as your child moves on to math operations that include regrouping.

During reading:

Point out the place value flats, rods, and blocks as you go through the story. Ask your child to count them to see if the total is the same as the number shown in the chart. For a challenge, cover up the place value chart and ask your child to predict how the number should be written, based on what the place value manipulatives show.

After reading:

Chow invents a zipometer that measures the speed at which he's traveling. What kind of math tool would your child like to invent? Ask them to draw their own invention and explain how it would work.

Celebrate your child's progress by placing a star sticker in the workbook.

True or False

Each of the pals is sharing a fun animal fact—but is it true or false? To find out, look at each math sentence. If the math fact is true, then so is the animal fact! You can use the boxes below the math sentences to help check the answers.

Giraffes only drink water every few days.

64 + 33 = 97

	+		=	
	+		=	

True ☐ **False** ☐

Crocodiles can hold their breath under water for more than an hour.

42 + 51 = 93

	+		=	
	+		=	

True ☐ **False** ☐

Rhinos never roll in mud.

26 + 72 = 88

	+		=	
	+		=	

True ☐ **False** ☐

Only the third fact is false. A rhinoceros rolls in mud to help protect its skin.

Your child has completed Step 3!
Celebrate by placing a star sticker in the workbook.

Launch

Regrouping to Add

Regrouping in a Place Value Chart

Sometimes when adding two-digit numbers, the sum in the ones place equals ten or more. You already know that each place can only have one digit, 0–9. This means the group of ones must be changed into a group of ten plus ones. The ones stay in the ones place, and the group of ten is moved to the tens place. This is called *regrouping*.

> Another name for regrouping is "carrying" because you "carry over" the extra digit.

Look at the problem 49 + 33 = ?

Start by adding the digits in the ones places.

Tens	Ones
4	9
+ 3	3
	12

The digits in the ones place have a value of 12.

Since 12 is a two-digit number, regroup it as 1 group of ten and 2 ones.

Tens	Ones
1	
4	9
+ 3	3
	2

The 2 ones stay in the ones place. The 1 group of ten is added to the other groups of ten in the tens place.

Add all the groups of ten together.

Tens	Ones
1	
4	9
+ 3	3
8	**2**

49 + 33 = **82**

Remind your child that in Math Fundamentals Step 1, they learned that numbers like 19 are ten plus nine more.

Hatch Match

Each dinosaur hatched from an egg, but which one? Solve each addition problem with regrouping, then match the answer to the dinosaur egg with the same number. The first one has been done for you.

Tens	Ones
1	
3	3
+ 5	8
9	1

Tens	Ones
2	9
+ 4	7

Tens	Ones
1	9
+ 2	6

Tens	Ones
6	6
+ 1	4

Tens	Ones
3	9
+ 3	3

It's important that your child fully understand place value to correctly perform regrouping. Remember, it's always fine to go back and review lessons as needed.

Launch

Regrouping with Place Value Blocks

Don't Forget!

Flat = 100 Rod = 10 Block = 1

Another way to look at regrouping is with place value blocks.

Remember, 1 tens rod = 10 ones blocks.

Show the numbers in the problem using place value blocks. 46 + 27 = ?	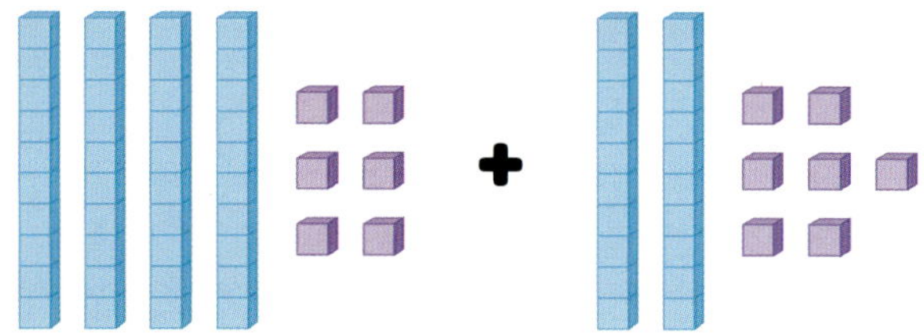	
Add the ones together. 6 + 7 = 13 Regroup the ones into a tens rod and ones blocks.	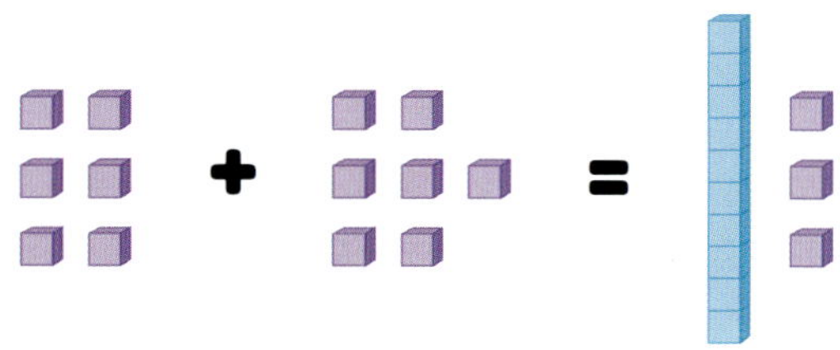	Tens: **1** Ones: **3**
Add the tens together, including the tens rod that was regrouped from the ones. 40 + 20 + 10 = 70	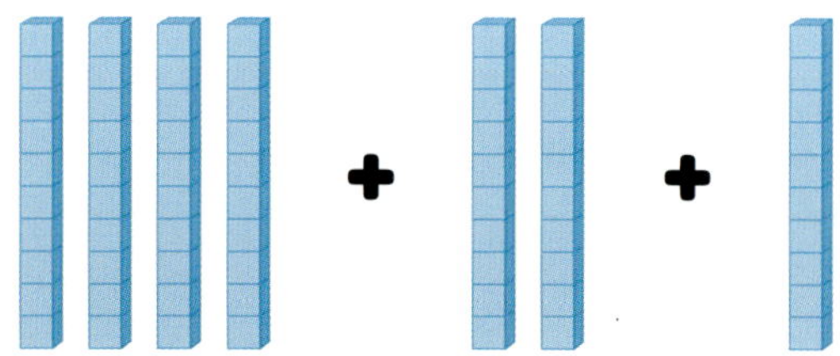	Tens: **7** Ones:
Add the tens and ones together.	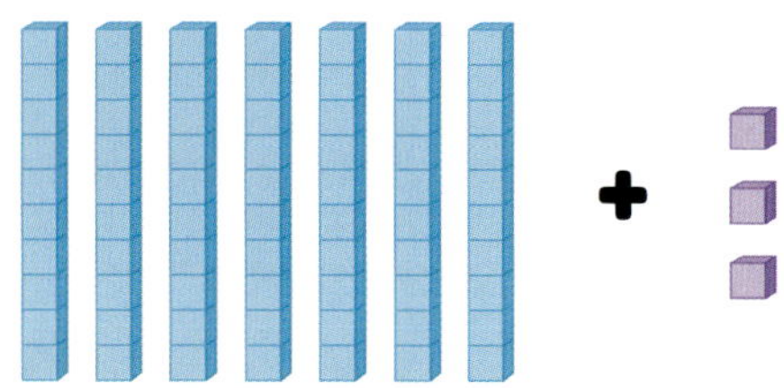	Tens: **7** Ones: **3**

In Step 11 of Math Fundamentals, your child learned to add place value blocks without regrouping. The steps above can help them see how 10 ones blocks can be regrouped into 1 tens rod.

Block by Block

Look at each group of place value rods and blocks. Write the number under each set.

Add the groups together, color the sum, then write the number. The first one has been done for you.

Hint: Circle the ones blocks first to count a set of 10. Color a tens rod in the sum, then color the remaining ones blocks in the sum.

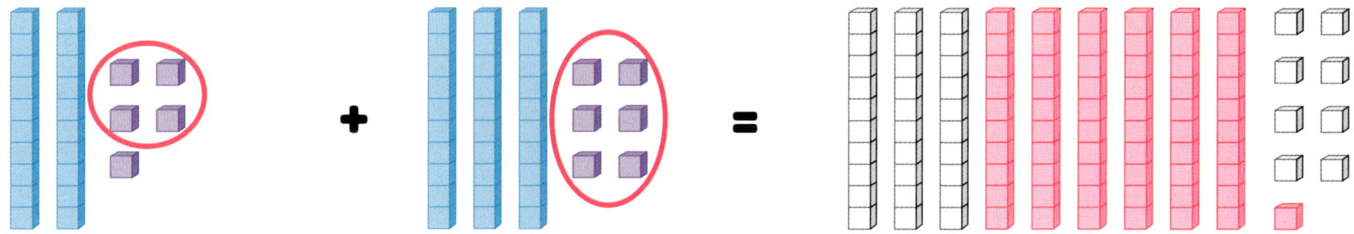

25 + 36 = 61

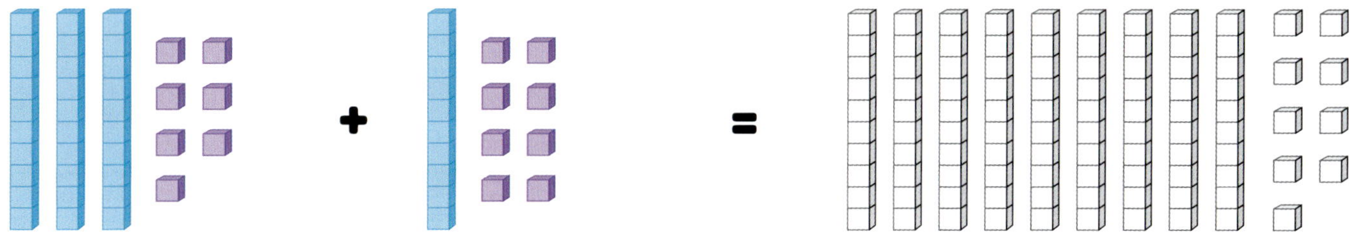

___ + ___ = ___

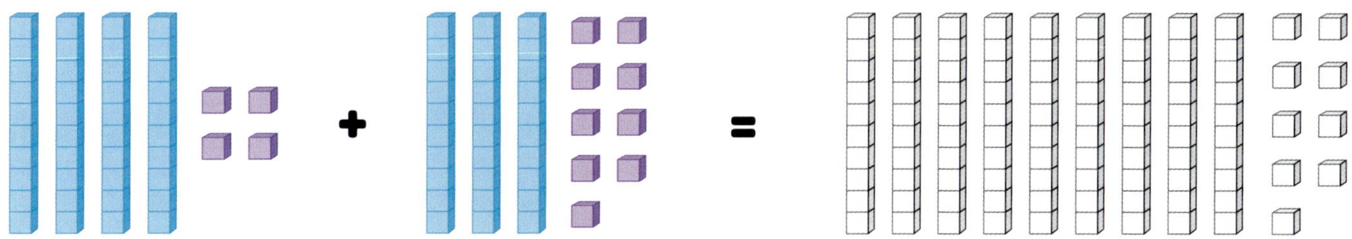

___ + ___ = ___

Launch

Word Problems with Regrouping

Quick Review!

Follow the same steps you learned earlier to solve word problems—just remember to regroup when you're adding!

Read the problem and search for key words and information.
Set up a horizontal math number sentence, then switch it to be vertical. When regrouping, remember to write the group of 10 at the top of the tens column to remind you to add it to the other tens.
Answer the word problem question in a complete sentence.

Tali baked muffins for her pals. She made **thirty-six** carrot muffins and **thirty-six** banana muffins. How many muffins did she bake **all together**?

36 + 36 = ?

$$\begin{array}{r} \scriptstyle 1 \\ 36 \\ +\,36 \\ \hline 72 \end{array}$$

You Did it!

Tali baked **72** muffins in all.

Reverse!

Write a word problem for the following equation: **56 + 35 = ?** Then solve it!

____

____

Writing the full answer to the question helps your child develop the habit of labeling quantities. Since mathematics is used for measuring quantities, it is necessary to label units of measure.

In a Few Words

Read the story in each word problem below. Then:

- Circle the numbers you need to work with.
- Underline the words that tell what operation to perform.
- Write each number sentence horizontally, and then vertically.
- Solve the problem and write the answer to the question. The first one has been done for you.

Story	Horizontal	Vertical	Answer
Batu made sixty-seven mud pies. Kazoo made fourteen mud pies. How many mud pies in all?	67 + 14 = ?	1 67 + 14 81	There are 81 mud pies in all.
Tali scored twenty-three points in the game. Chow scored eighteen. How many points did both players score all together?			
It's clean-up day. The pals picked up nineteen bags of trash at the park. Then they cleaned up seventeen bags of trash by the river. How many bags of trash in all?			

Part-Part-Whole Strategy with Two-digit Numbers

Break It Down!

You have used the part-part-whole method with single-digit numbers. You can also use the part-part-whole method to break apart double-digit numbers to make equations easier to understand and solve.

Break down one number or both numbers into parts that are sets of tens and ones. In this problem break the whole, **31**, into two parts, **30** and **1**.	**31 + 62 = ?** **30 + 1 + 62 = ?**
Think of **30** as 3 tens. Add them to **62** by skip counting. Start at **62**, and count 72, 82, **92**.	**30 + 62 + 1 = ?** **92 + 1 = ?**
Add the ones to find the sum.	**92 + 1 = 93**

There are different ways to break numbers down so they are easier to add in your head. Remember earlier strategies such as doubles, making 10, and doubles + 1 or doubles – 1.

Be Strategic

Look at the way each problem is broken down. Match the strategy used for each one.

54 + 36 = ?
50 + 4 + 30 + 6 = ?
50 + 30 + 4 + 6 = ?
50 + 30 + 10 = 90

43 + 23 = ?
40 + 3 + 20 + 3 = ?
40 + 20 + 3 + 3 = ?
40 + 20 + 6 = 66

24 + 15 = ?
20 + 4 + 10 + 5 =
20 + 10 + 4 + 5 =
20 + 10 + 9 = 39

doubles   doubles + 1   make 10

These strategies were reviewed in Step 2 of Elementary Math.

Missing Parts

Each problem is broken down into parts to solve. But some of the parts are missing! Look at each equation and fill in the missing numbers. The first one has been done for you.

41 + 36 = ?	37 + 53 = ?	85 + 15 = ?
40 + 1 + 30 + 6 = ?	30 + ___ + ___ + 3 = ?	85 + ___ + 5 = ?
40 + 30 + 1 + 6 = ?	30 + ___ + 7 + ___ = ?	85 + ___ + 10 = ?
40 + 30 + 7 = 77	30 + 50 + ___ = 90	90 + 10 = ___

Smashing Success

Break down the problems below any way you like to solve them.

52 + 48 = ?	16 + 71 = ?	38 + 21 = ?

Breaking down numbers is a useful strategy for doing math "in your head" without pencil and paper. It's often called *mental math*.

A Picnic Place

Tali is preparing for a picnic. Where will she meet her friends? Solve the word problems, then color in the matching numbers on the next page to see where Tali is meeting her pals.

First, to get ready for the picnic, Tali needs twenty-five forks and thirty-five spoons. How many utensils does she need in all?

Tens	Ones
+	

Answer: ______

Second, Tali makes thirty-two pickle sandwiches and nineteen peanut butter sandwiches. How many sandwiches does she make in all?

Tens	Ones
+	

Answer: ______

Third, Tali adds forty-six apple slices and thirty-seven orange slices to her picnic basket. How many slices of fruit all together?

Tens	Ones
+	

Answer: ______

Follow the Path

Solve the first problem on the previous page, then color the same numbers and follow the path to see where Tali will have her picnic.

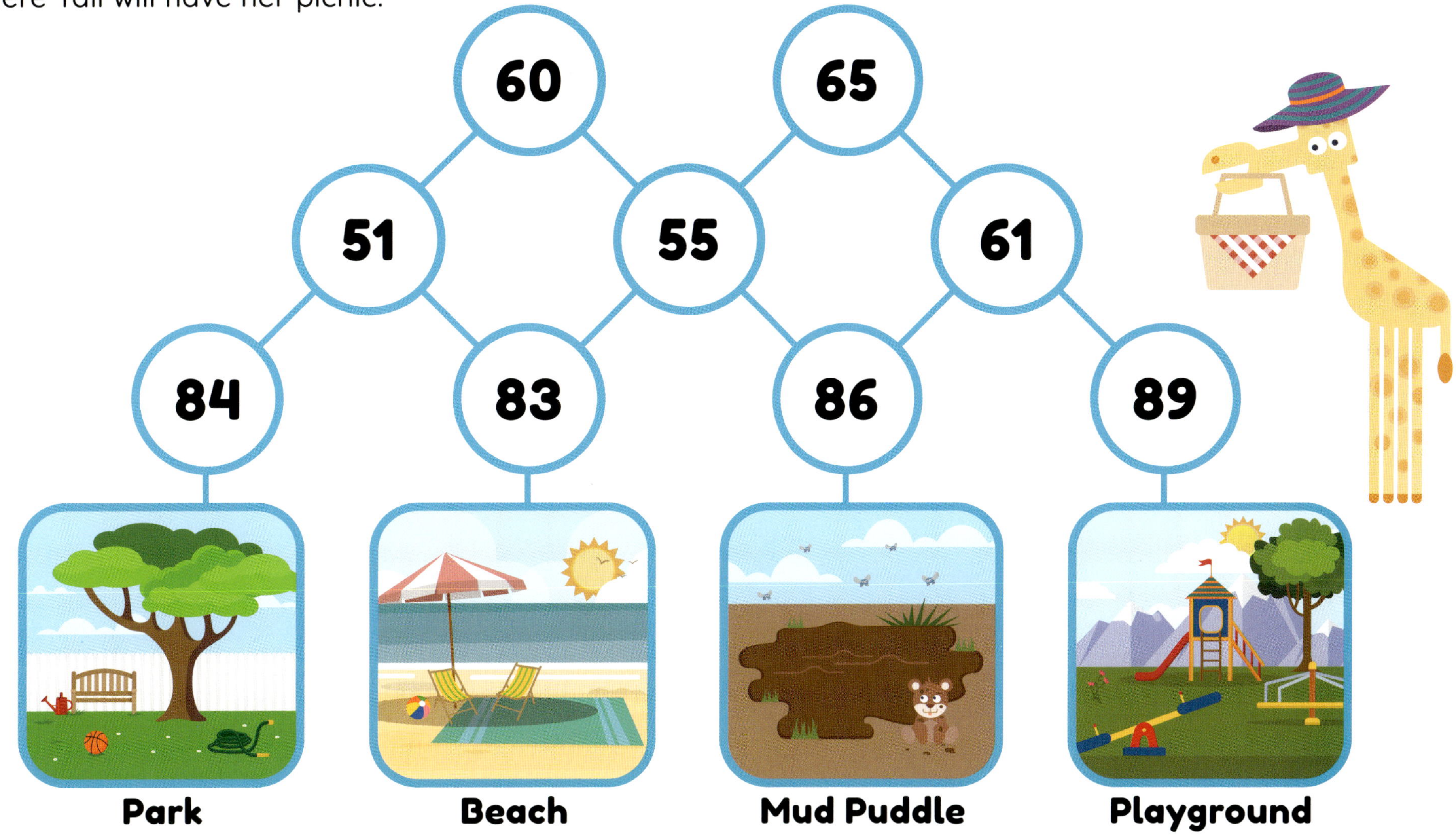

BINGO Spin

Go for a Spin

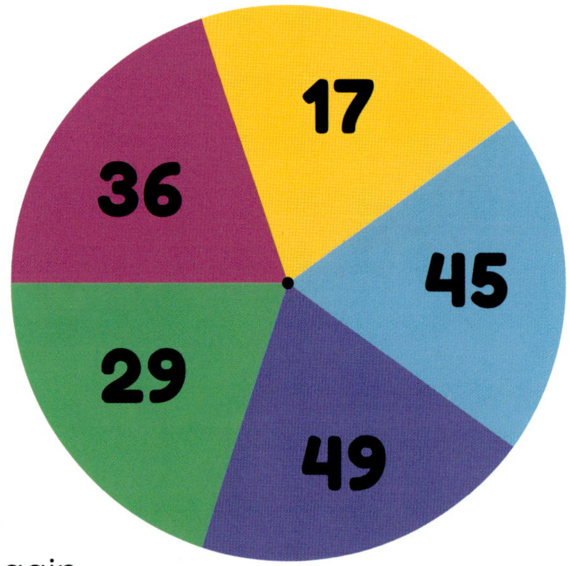

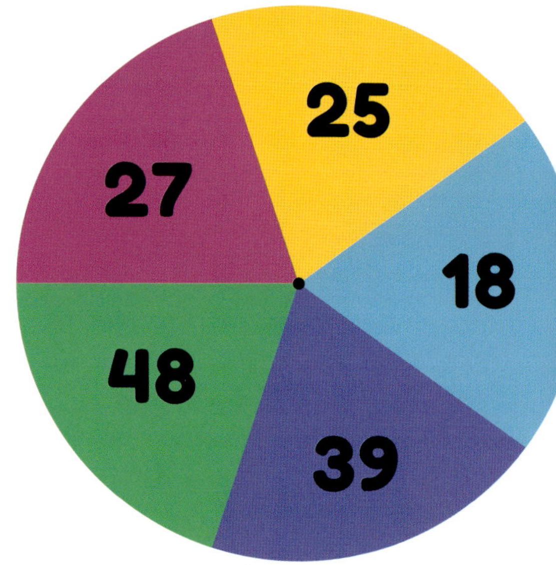

1. Spin Spinner 1 and write the number in the top box.
2. Spin Spinner 2 and write the number in the box below it.
3. Add up the two numbers and write the sum.
4. Place an **X** over that number on the BINGO cards.
5. Keep spinning and adding. If you spin a sum you have already made, spin again.
6. The first card to get five numbers in a row across, down, or diagonally wins!

You can play alone or with a friend. If you like, play again with any extra boxes or scrap paper to help you add.

Tens	Ones
+	

Tens	Ones
+	

Tens	Ones
+	

Tens	Ones
+	

Tens	Ones
+	

Materials needed: sharpened pencil, paper clip

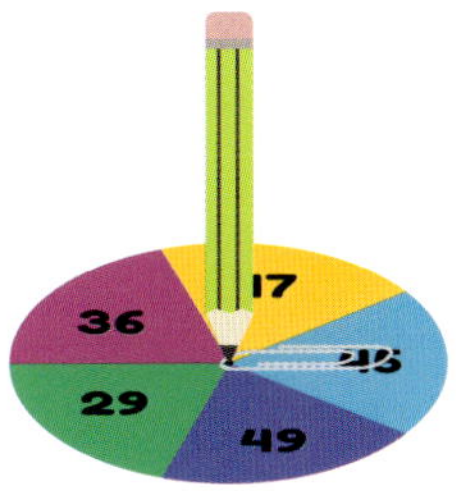

How to Make the Spinner: Place one end of the paper clip in the center of the circle. Put the pencil tip inside the paper clip on the center dot. Now flick the paper clip so it spins around the pencil.

B	I	N	G	O
Free	84	75	54	61
56	77	68	47	54
76	97	Free	67	74
72	93	88	63	70
44	65	42	35	Free

B	I	N	G	O
77	47	54	97	Free
75	56	76	67	74
54	61	Free	72	88
63	70	93	44	84
Free	65	42	35	97

Tens	Ones
+	

Tens	Ones
+	

Tens	Ones
+	

Tens	Ones
+	

Tens	Ones
+	

Read — A Visit with Madam Addison

Your child is ready to read *A Visit with Madam Addison*, a whimsical story that features addition problems wit regrouping. To reinforce this concept, color-coded numbers help your child follow the numbers being added. This storybook will help them to become more confident solving addition problems that require regrouping.

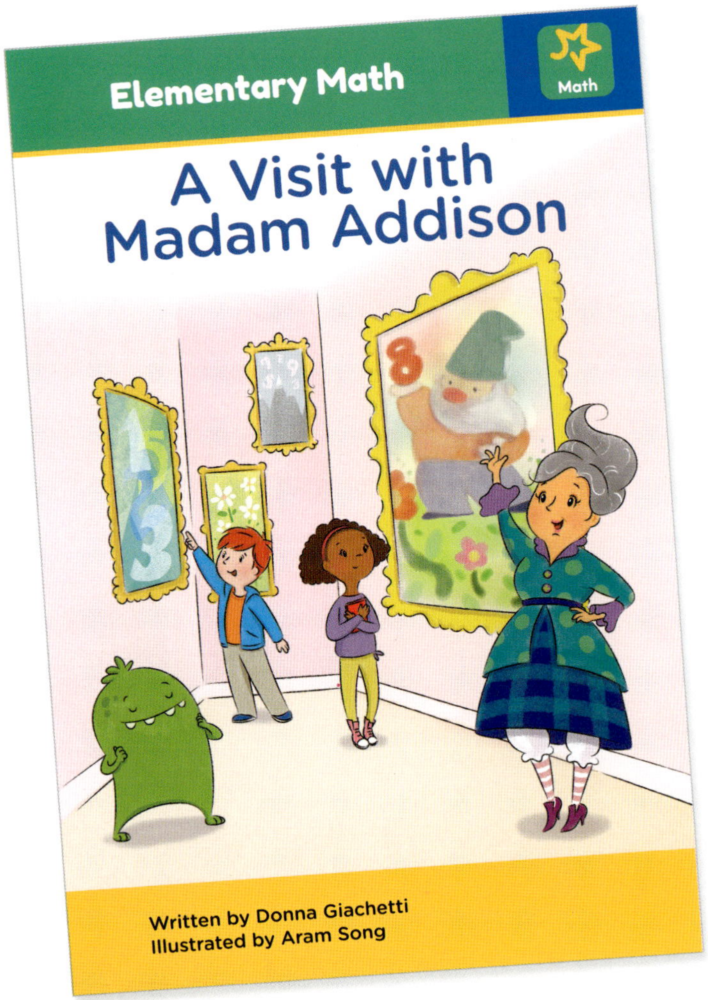

During reading:

Occasionally cover up the answer to one of the addition problems presented in the book. Ask your child to explain the steps that are required to solve it.

Discuss the idea of “carrying over” or “regrouping.” Make sure your child understands that the digits represent the number in that group, and that each place can only hold a single digit.

After reading:

Madam Addison’s favorite hobby is adding things together. Ask your child if they can imagine taking up a hobby like Madam Addison’s. What sort of hobby might they enjoy?

Madam Addison has a very unusual helper, Swifty. Have your child draw a picture of the kind of creature they would like to befriend. What would they name it?

Celebrate your child’s progress by placing a star sticker in the workbook.

Pick a Pop!

To find out the flavor of the ice pops, match each number sentence to a sum in the middle pop, then color that section of pop the same color. The first one has been done for you.

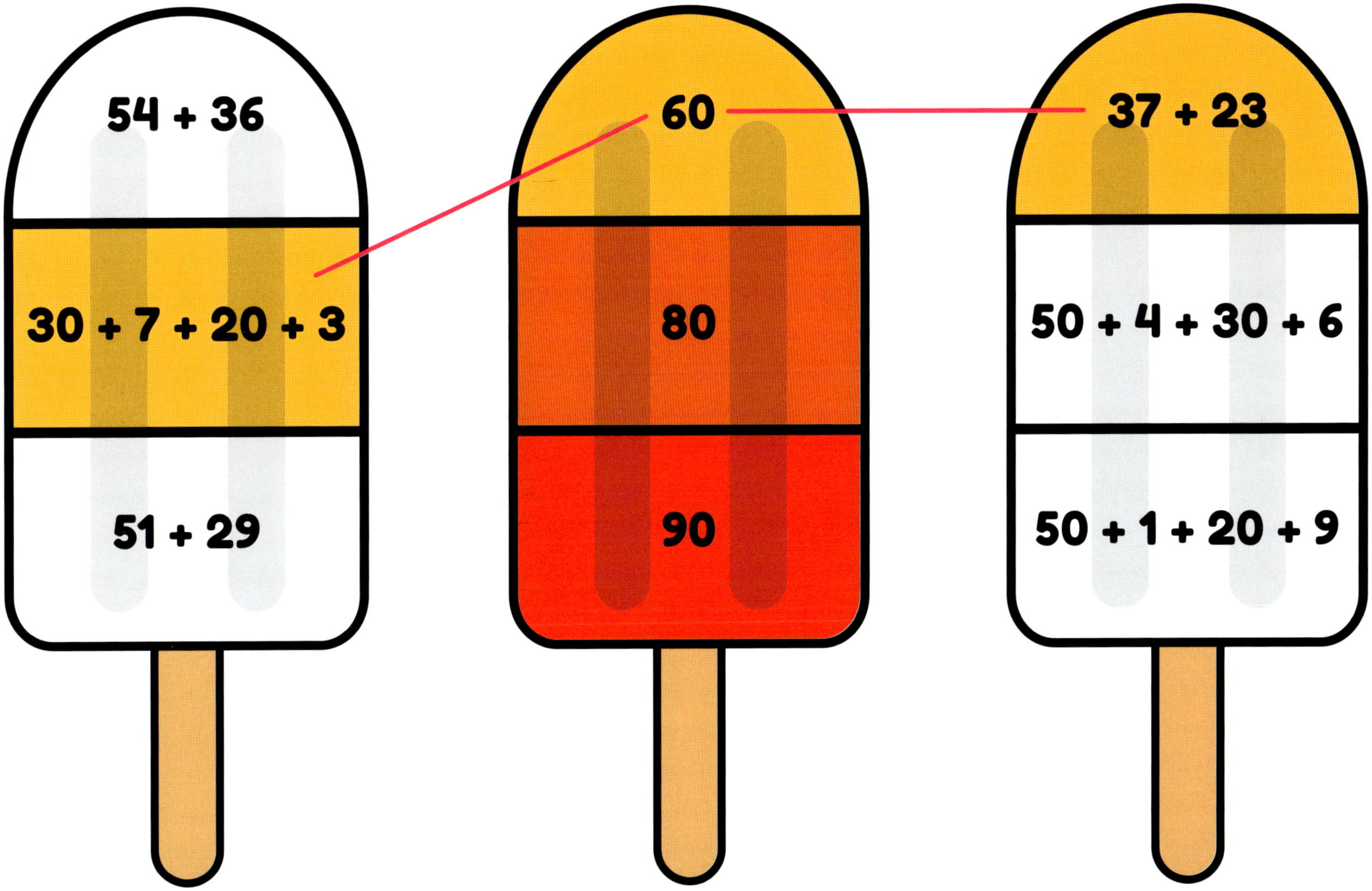

Hooked on Math

It's Time to Celebrate!

____

has completed Elementary Math Steps 3 & 4 and is now ready to move on to the Elementary Math Steps 5 & 6 Workbook!

SKILLS MASTERED

- Addition word problems with and without regrouping
- Addition regrouping with a place value chart
- Addition regrouping with place value blocks
- Decomposing numbers to add

Don't forget to share your success and tag us on social media so we can celebrate with you!

Credits

Design Team
Julie Linden Frazier
Lindsey Frazier

Editorial Team
Donna Giachetti
Cynthia Stierle
Austin Feld

Illustrations
Dubit Limited

Educational Advisor
Lesley Freed, Curriculum Specialist

Published by Sandviks HOP, Inc.

Printed in China

Hooked on Math

Elementary Math

18037 (11/22)

Hooked on Math

Elementary Math

Steps 1 & 2

Workbook

Tips for Success

- Elementary Math focuses on understanding the importance of place value while working with numbers up to 1,000. It also introduces the key concept of regrouping and builds on the addition and subtraction skills introduced in Math Fundamentals.
- Work at a pace that's right for you and your child.
- Fit your session length to your child's attention span. It's okay to work with the App one day and this workbook the next.
- Before beginning, flip through the workbook and note any materials you might need. You should plan to have pencils, a pair of dice, and crayons ready.
- You'll see parent tips throughout this workbook. These tips are designed to answer questions you or your child might have about the learning.
- Repetition and review are essential to basic math fluency. Each workbook step builds on the ones before, so make sure your child has mastered the current lesson before moving forward.
- As you go through the workbook, encourage your child to explore the connection between numbers and the real world.
- Have fun with it!

Hooked on Math

Elementary Math

Steps 1 & 2

Workbook

This workbook and the included storybooks introduce elementary math skills.

Step 1: Number Properties
- Even and odd numbers
- Equal groups
- Repeated addition
- Arrays

Step 2: Strategies within 20
- Add in any order
- Doubles and near doubles
- Adding three numbers
- Part-part-whole number bonds

Hooked & Company

Contents

Lesson Overview

Welcome! This workbook provides supplemental activities that will solidify and expand the learning your child is enjoying with the Hooked on Math App. Follow the easy, step-by-step instructions to guide your child's exploration of numbers and operations. Every lesson helps your child build a strong foundation in elementary math skills while having fun!

1. Launch

Your child will be introduced to math concepts and skills through interactive lessons in the App on your mobile device or Windows or Mac computer. Download the App at:

hookedonmath.com/get-the-app

2. Discover

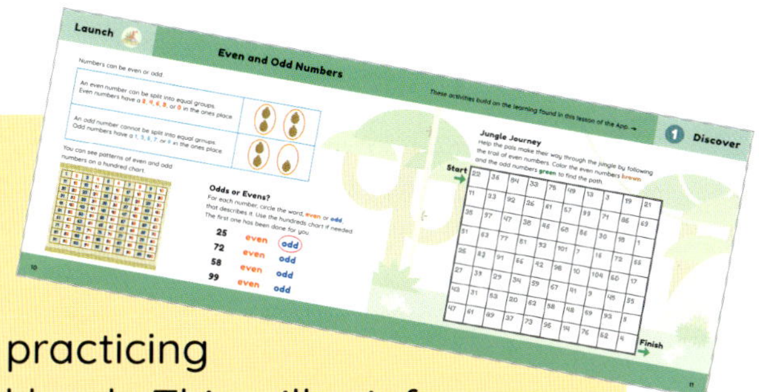

Next, your child will have fun exploring math concepts and practicing new skills in the workbook. This will reinforce the App lessons and provide a tangible application of the skills learned.

3. Play

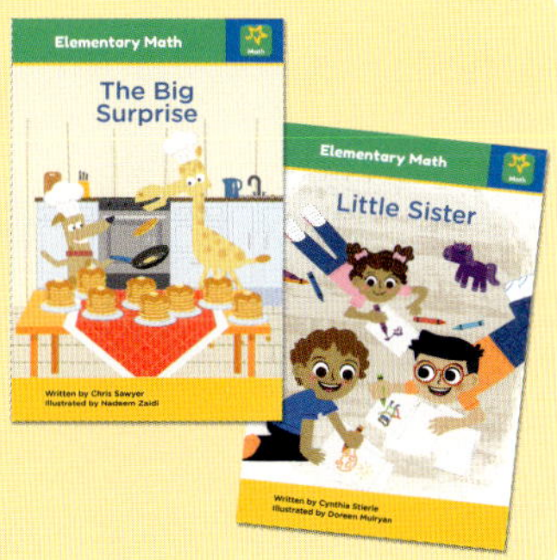

In addition to the workbook activities, your child will enjoy two math-based storybooks that will strengthen their learning and add to the fun!

4. Celebrate

Celebrate your child's progress by placing a sticker in the workbook as learning milestones are reached. This also provides a helpful marker to remember where to begin next time.

Special notes are included in this space to provide additional context related to what is being taught.

City Lights

Circle the number below each building that is the best estimate for how many windows are shown. Then count the number of windows to check your estimate. The first one has been done for you.

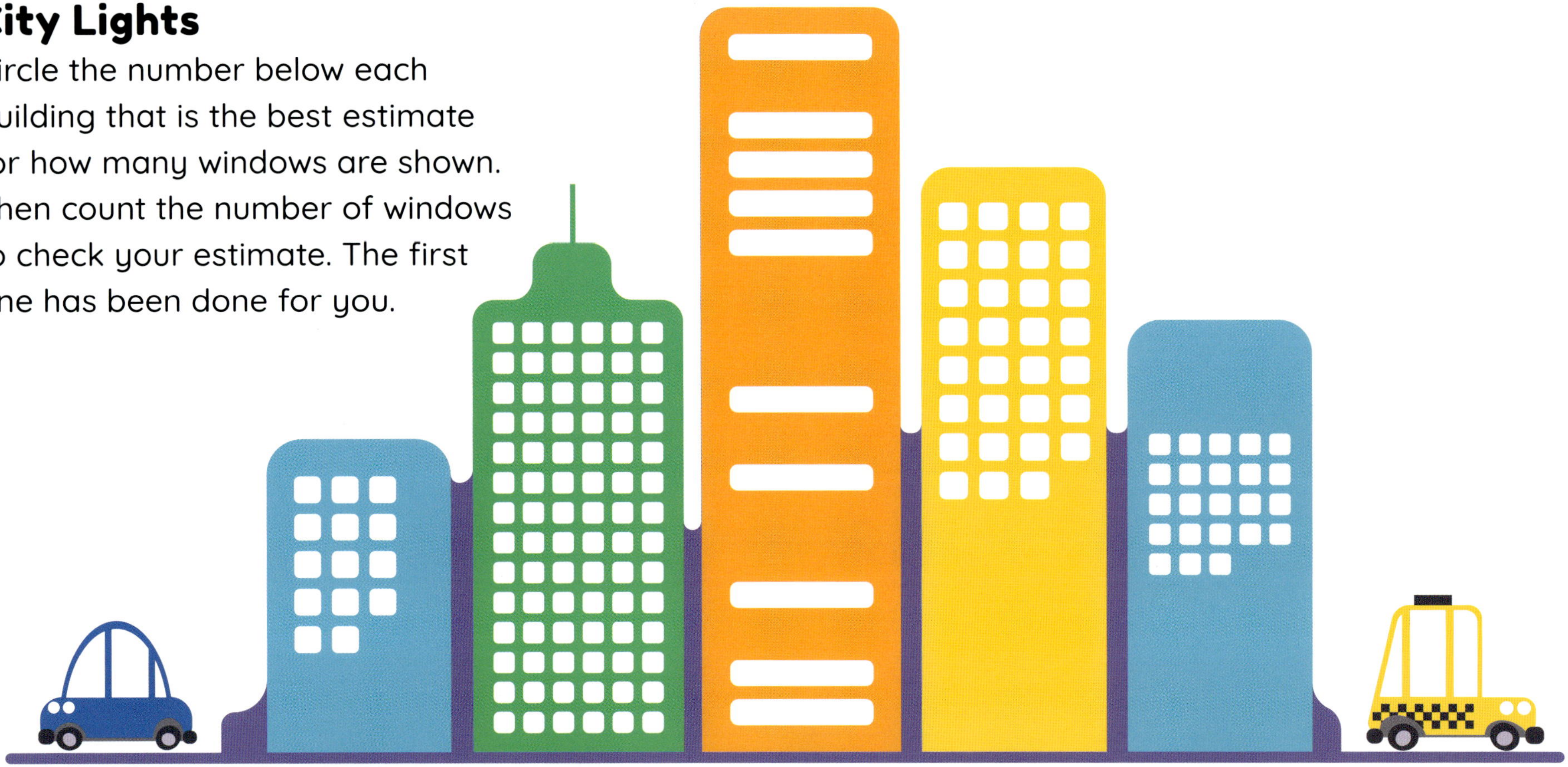

60	30	40	80	100
5	80	25	30	25
(15)	200	11	100	50
14	___	___	___	___

Draw a line to match each number representation to the same number inside the coconut.
The first one has been done for you.

1	2	3	4	5	6	7	8	9	10
11	12	13	14	15	16	17	18	19	20
21	22	23	24	25	26	27	28	29	30
31	32	33	34	35	36	37	38	39	40
41	42	43	44	45	46	47	48	49	50
51	52	53	54	55	56	57	58	59	60
61	62	63	64	65	66	67	68	69	70
71	72	73	74	75	76	77	78	79	80
81	82	83	84	85	86	87	88	89	90
91	92	93	94	95	96	97	98	99	100

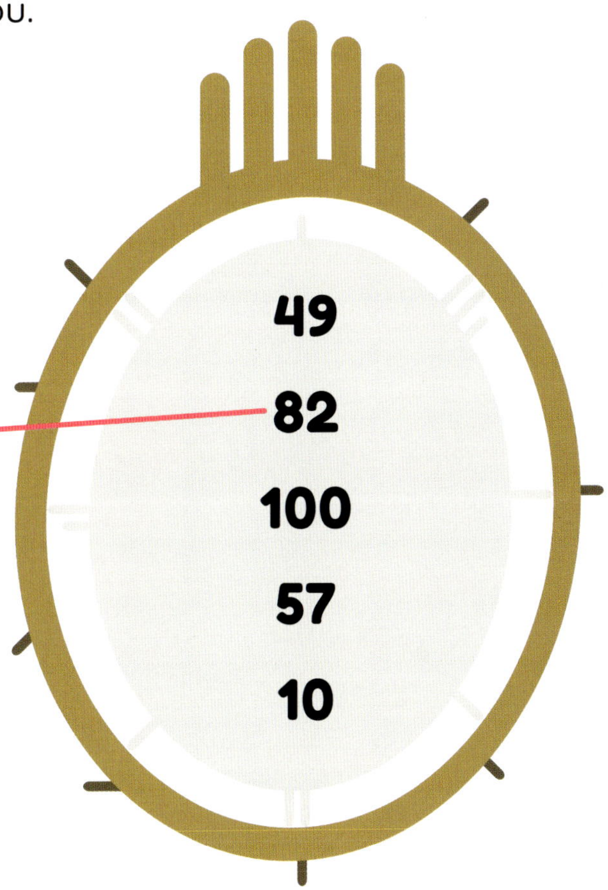

Ten

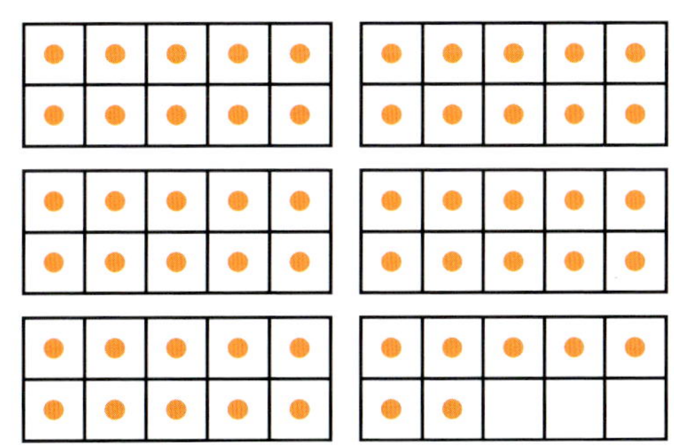

Place Value to 120

How Many?

Look at the place value flats, rods, and blocks on these pages. Write how many each one represents on the line below the picture. Then add them together and write the sum. The first one has been done for you.

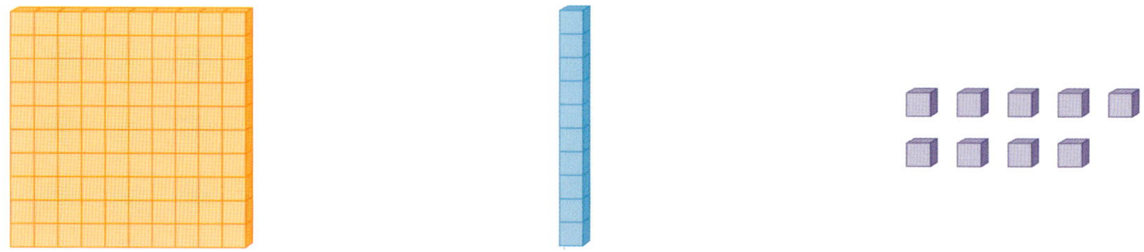

100 + 10 + 9 = 119

____ + ____ + ____ = ____

____ + ____ + ____ = ____

Zero at Work:
See how 0 works as a placeholder? Remember, if there is a hundreds flat, but no tens rods or ones blocks, write in 0 as a placeholder for tens and/or ones.

___ + ___ + ___ = ___

___ + ___ + ___ = ___

___ + ___ + ___ = ___

Piles of Pancakes

Batu is making pancakes for his friends. Add the numbers together to see how many pancakes he makes for each pal. Then circle the pal with the most pancakes. The first one has been done for you.

Tens	Ones
3	4
+ 1	3
4	7

How many pancakes are on the plate?
47

Tens	Ones
8	1
+ 1	7

How many pancakes are on the plate? ____

Tens	Ones
2	6
+ 2	3

How many pancakes are on the plate? ____

Tens	Ones
4	1
+ 1	1

How many pancakes are on the plate? ____

Tens	Ones
1	5
+ 5	2

How many pancakes are on the plate? ____

Tens and Ones Subtraction

Review

All Done!

The pals were hungry! Notice the number of pancakes each pal ate. Subtract to find out who has the fewest pancakes left. Circle that pal.

Tens	Ones
4	7
- 2	1

How many pancakes are on the plate?

____

Tens	Ones
9	8
- 8	4

How many pancakes are on the plate?

____

Tens	Ones
4	9
- 3	2

How many pancakes are on the plate?

____

Tens	Ones
5	2
- 2	1

How many pancakes are on the plate?

____

Tens	Ones
6	7
- 4	3

How many pancakes are on the plate?

____

Celebrate finishing the review with a star sticker.

Launch

Even and Odd Numbers

Numbers can be even or odd.

An *even* number can be split into equal groups. Even numbers have a **2**, **4**, **6**, **8**, or **0** in the ones place.	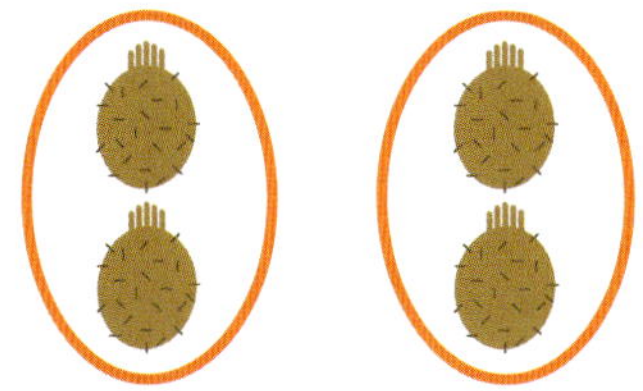
An *odd* number cannot be split into equal groups. Odd numbers have a **1**, **3**, **5**, **7**, or **9** in the ones place.	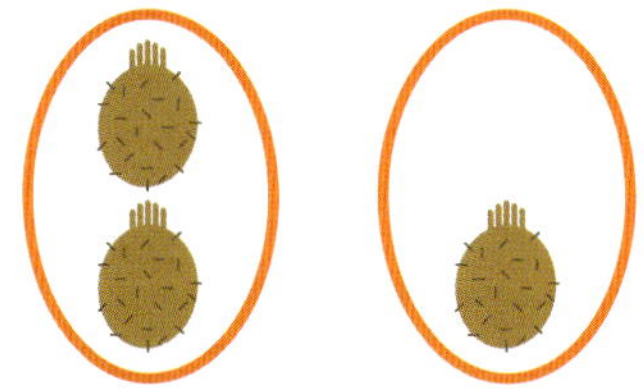

You can see patterns of even and odd numbers on a hundred chart.

1	2	3	4	5	6	7	8	9	10
11	12	13	14	15	16	17	18	19	20
21	22	23	24	25	26	27	28	29	30
31	32	33	34	35	36	37	38	39	40
41	42	43	44	45	46	47	48	49	50
51	52	53	54	55	56	57	58	59	60
61	62	63	64	65	66	67	68	69	70
71	72	73	74	75	76	77	78	79	80
81	82	83	84	85	86	87	88	89	90
91	92	93	94	95	96	97	98	99	100

Odds or Evens?

For each number, circle the word, **even** or **odd**, that describes it. Use the hundreds chart if needed. The first one has been done for you.

25	**even**	**odd** (circled)
72	**even**	**odd**
58	**even**	**odd**
99	**even**	**odd**

Jungle Journey

Help the pals make their way through the jungle by following the trail of even numbers. Color the even numbers **brown** and the odd numbers **green** to find the path.

Start →

22	36	84	33	75	49	13	3	19	21
11	23	92	26	61	57	99	71	85	69
35	97	47	38	46	68	86	30	18	1
51	63	77	81	93	101	7	15	72	65
25	83	91	56	42	98	10	104	50	17
27	39	29	34	59	67	41	9	45	55
43	31	53	20	62	58	48	59	93	5
47	61	89	37	73	95	14	76	52	4

Finish →

Launch

Equal Groups

Groups are equal when there are the same number of items in each group.

The number 6 can be split into two groups with 3 in each group.	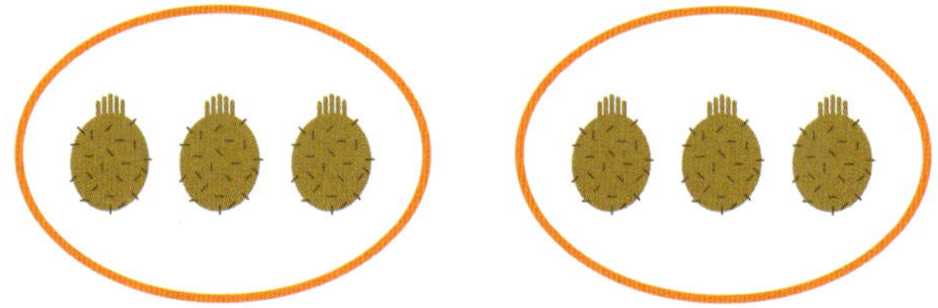
The number 6 can also be split into three groups with 2 in each group.	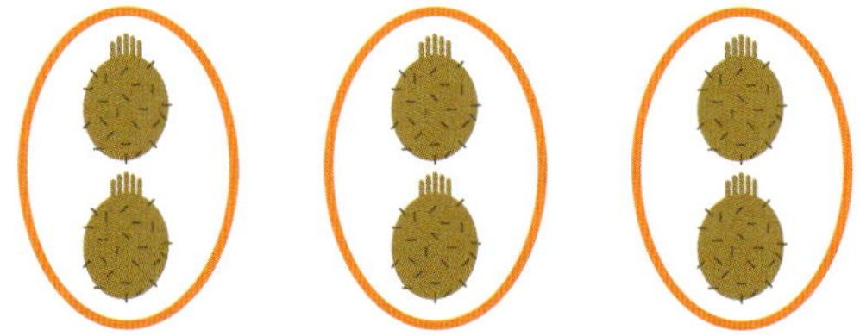

If each group has the same number of objects, it is an equal group.

Banana Splits

Read each description, then split the bananas into equal groups by circling them. The first one has been done for you.

4 equal groups of 2

3 equal groups of 5

6 equal groups of 3

Turtles and Eggs

Look at the groups in each row. Count the groups and write how many in the first column. Then count how many in each group and write it in the second column. The first one has been done for you.

	Number of Groups	Number in Each Group
	5	2
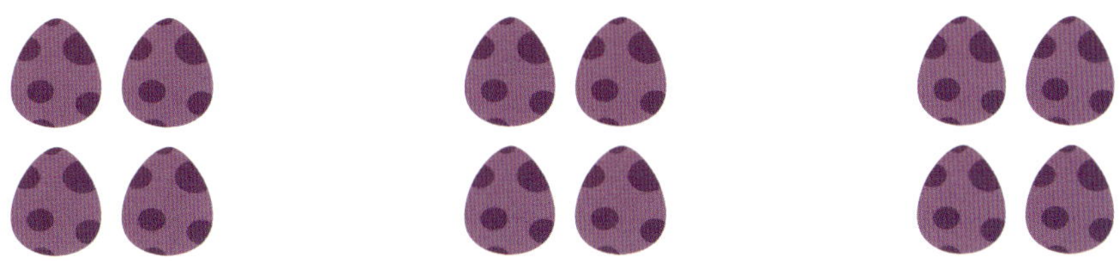		

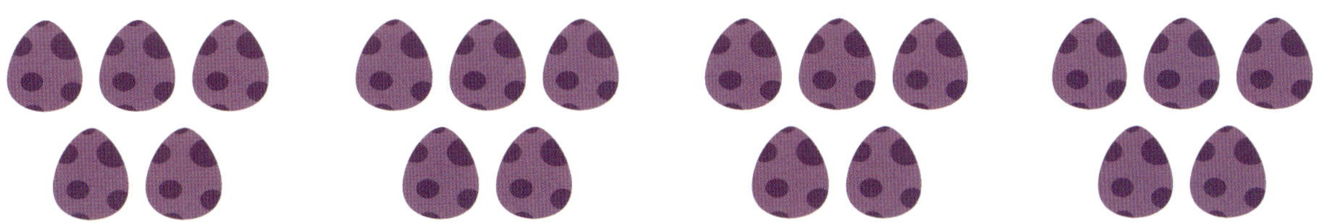		

Repeated Addition

Repeated addition is adding equal groups together. It is a strategy that can help you find the total number in a larger group. You can try using doubles facts or skip counting to find the sum.

A Leg Up

There are 4 turtles. Each turtle has 4 legs. How many legs in all? Instead of counting each leg, use repeated addition.

4 + 4 + 4 + 4 = ?

8 + 8 = 16

Take Note

Look at each picture.

- Write how many music notes are in each group and how many groups.
- Write the repeated addition sentence and find the sum.

The first one has been done for you.

There are 6 groups with 2 in each group.

2 + 2 + 2 + 2 + 2 + 2 = 12

There are ___ groups with ___ in each group.

____

There are ___ groups with ___ in each group.

____

Bird Watching

Each flock of birds is in an equal group. Count the birds and write the repeated addition sentence. Then find the sum. The first one has been done for you.

4 + 4 + 4 = 12

____ + ____ + ____ = ____

____ + ____ + ____ + ____ = ____

Arrays

An *array* is a group of objects, pictures, or numbers that shows equal groups arranged in rows and columns.

Rows are horizontal, or across.

5 + 5 + 5 + 5 = 20

Columns are vertical, or up and down.

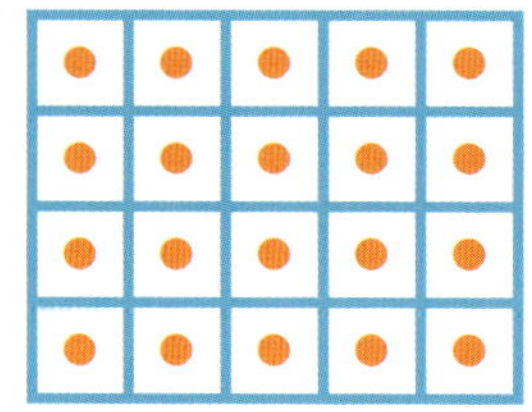

4 + 4 + 4 + 4 + 4 = 20

You can use repeated addition to find the number of objects in an array: add the number of objects in each row by the number of rows (**5 + 5 + 5 + 5 = 20**) or add the number of objects in each column by the number of columns (**4 + 4 + 4 + 4 + 4 = 20**). Either way, the sum will be the same!

Arrange an Array

Draw a dot in each box to show the number sentence. The first one has been done for you. (**Hint:** You may not need to fill all the boxes.)

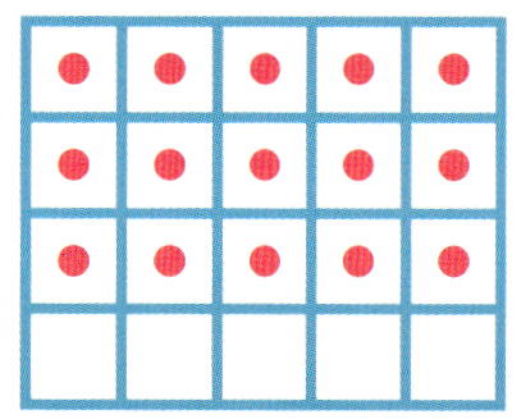

5 + 5 + 5 = 15

3 + 3 + 3 + 3 = 12

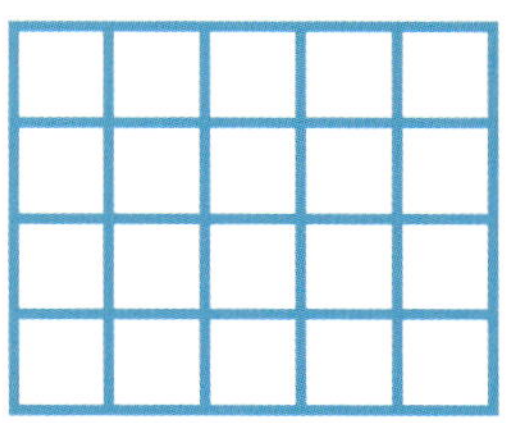

4 + 4 + 4 = 12

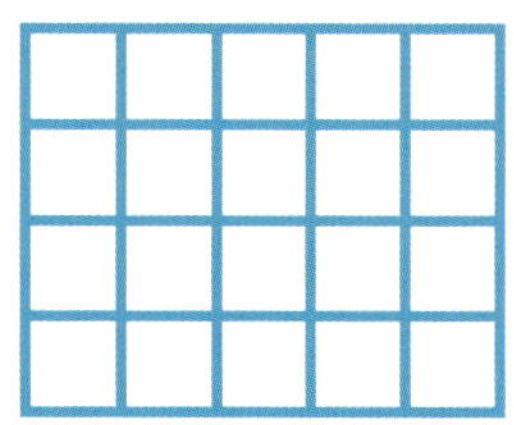

4 + 4 + 4 + 4 = 16

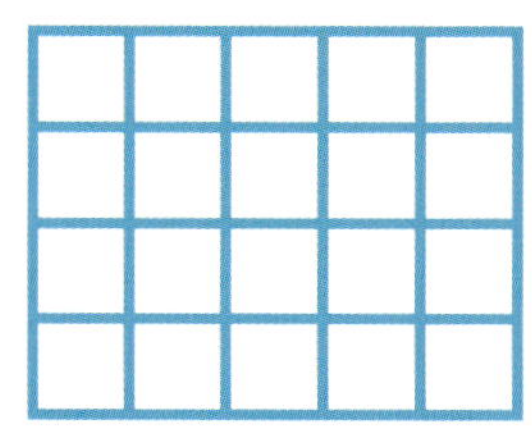

2 + 2 + 2 + 2 = 8

Understanding arrays and repeated addition will help when learning multiplication and division.

Lots of Dots

Draw a line to match each repeated addition sentence to one of the arrays below.

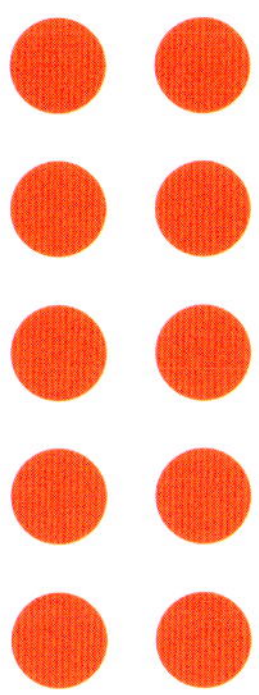

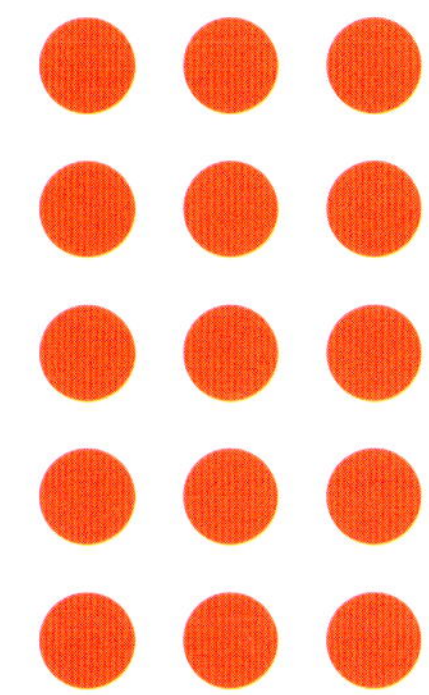

$2 + 2 + 2 + 2 = 8$

$6 + 6 + 6 = 18$

$5 + 5 + 5 = 15$

$2 + 2 + 2 + 2 + 2 = 10$

$5 + 5 + 5 + 5 = 20$

$4 + 4 + 4 + 4 = 16$

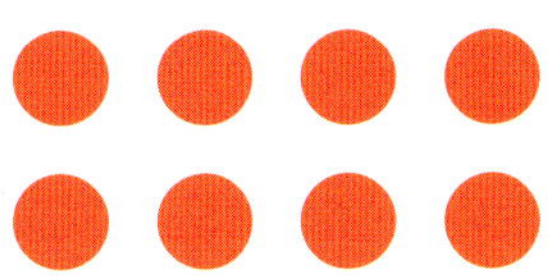

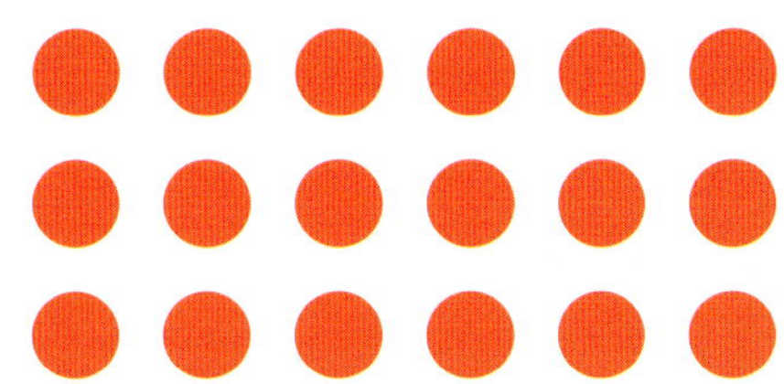

Donut Delight

Write two different repeated addition sentences for each array. The first one has been done for you.

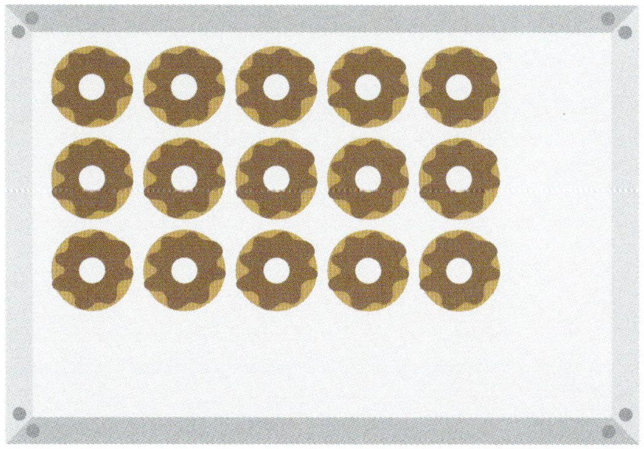

3 + 3 + 3 + 3 + 3 = 15

5 + 5 + 5 = 15

Repeated addition sentences for rectangular arrays can be composed in two ways: as 2 + 2 + 2 + 2 = 8 and 4 + 4 = 8.

Odds or Evens?

Play

Tic-Tac-Toe Time

- Find the sum for each addition problem.
- Circle the sums that are even numbers.
- Draw an **X** over the sums that are odd numbers.

Which numbers win?

3 + 3 + 3 =	6 + 6 + 6 + 6 =	4 + 4 + 4 =
5 + 5 + 5 =	5 + 5 + 5 + 5 =	1 + 1 + 1 =
7 + 7 + 7 =	9 + 9 + 9 =	2 + 2 + 2 + 2 =

4 + 4 + 4 + 4 =	10 + 10 + 10 =	1 + 1 + 1 + 1 + 1 =
11 + 11 + 11 =	7 + 7 + 7 + 7 =	2 + 2 + 2 =
7 + 7 =	3 + 3 + 3 + 3 =	5 + 5 + 5 + 5 + 5 =

Riddle: Why didn't the swimmer like to race in lanes 1, 3, or 5?

Answer: He didn't like the odds.

Read

The Big Surprise

Your child is ready to read *The Big Surprise*. This storybook features arrays and repeated addition. Remembe that an array is a visual representation of objects that shows equal groups arranged in rows and columns. After the array is created, repeated addition is used to find out how many objects there are in all.

During reading:

Point out the arrays to your child. Together, count the number of columns and rows. How many equal groups are there? How many in each group? Help your child to recognize that those numbers are represented in the accompanying math equation—either as actual numerals, or as the number of times a numeral is repeatedly added.

After reading:

Have your child create their own arrays. They can place small objects into empty muffin tins or egg cartons or draw arrays using paper and crayons. When they finish creating the array, ask them to write the addition sentence that matches it and find the sum.

Kazoo and Tali cooked 48 pancakes (and many other goodies) in *The Big Surprise*. After reading the story, why not make a batch of pancakes with your child?

Celebrate your child's progress by placing a star sticker in the workbook.

Hooray for the Array

Fill in the blanks and draw pictures in the array to tell a story.

Example

Kazoo (my dog) (name) **has** **2** (digit between 1-4) **equal groups of** **bones** (small objects).

There are **3** (digit between 1-4) **bones** (same objects) **in each group. That**

means that **Kazoo** (name) **has** **6** (total number) **bones** (same objects).

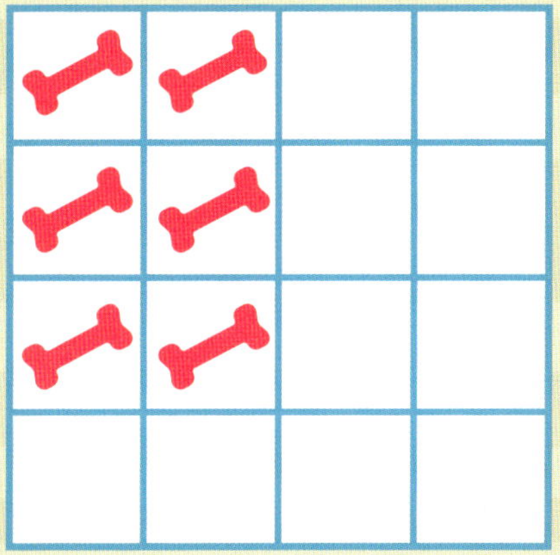

Activity

____________ (name) **has** **____** (digit between 1-4) **equal groups of** **____________** (small objects).

There are **____** (digit between 1-4) **____________** (same objects) **in each group. That**

means that **____________** (name) **has** **____** (total number) **____________** (same objects).

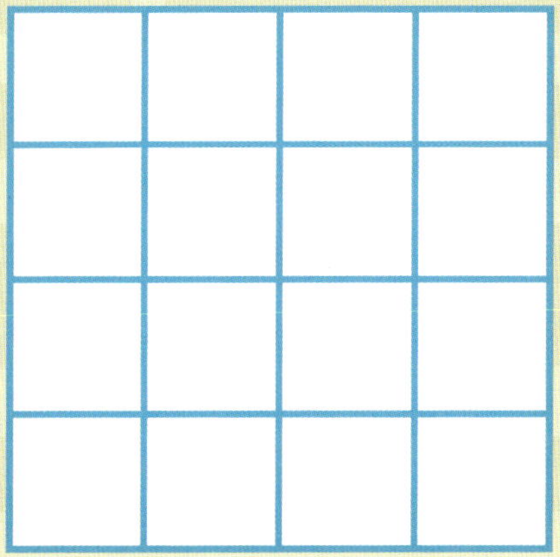

Your child has completed Step 1!
Celebrate by placing a star sticker in the workbook.

Add in Any Order

Follow the Rules

In math, a *property* is a rule that must be followed.

The *commutative property of addition* states that when adding, numbers can swap places and the answer, or sum, will always be the same. This is true even when more than two numbers are being added.	6 + 5 + 4 = 15 4 + 6 + 5 = 15 5 + 4 + 6 = 15

The commutative property does NOT work with subtraction.

Missing Numbers

Look at the first addition sentence in each row. Use the commutative property of addition to fill in the missing numbers in the remaining sentences. The first one has been done for you.

12 + 3 + 6 = 21	3 + 6 + 12 = 21	6 + 12 + 3 = 21
15 + 7 + 8 = 30	7 + 8 + ___ = 30	___ + 15 + ___ = 30
11 + 19 + 26 = 56	___ + 19 + 11 = 56	19 + ___ + ___ = 56
55 + 2 + 33 = 90	33 + ___ + 2 = 90	___ + ___ + 55 = 90

Re-Leaf!

Draw a line to match the number sentence on each leaf to its sum on the tree trunk. Use the commutative property of addition to help you. The first one has been done for you.

Doubles and Near Doubles

Doubles: Quick Check

Doubles facts are number sentences that show the same number added to itself. Do you know your doubles facts? Do a quick check by finding the sums for each fact below.

1 + 1 = ___	2 + 2 = ___	3 + 3 = ___	4 + 4 = ___	5 + 5 = ___
6 + 6 = ___	7 + 7 = ___	8 + 8 = ___	9 + 9 = ___	10 + 10 = ___

Near Doubles Strategy

When you have a number sentence that is almost a double, you can use the doubles + 1 strategy.

Notice the number 9 is 1 more than 8.	8 + 9 
Write the sentence as a doubles + 1.	8 + 8 + 1 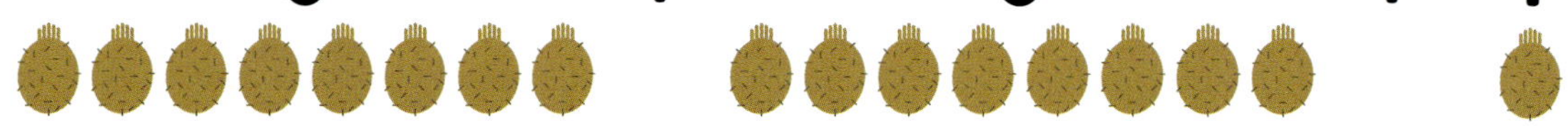
Add the doubles together.	16 + 1 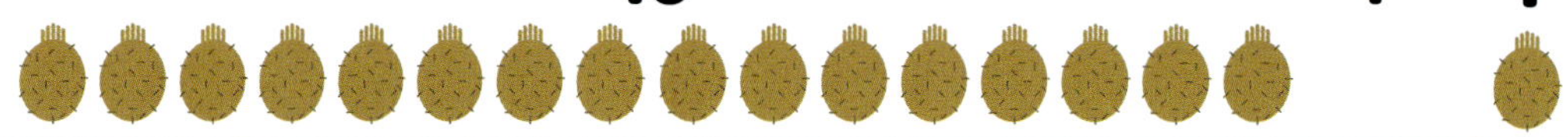
The sum is 1 more than the doubles fact.	17 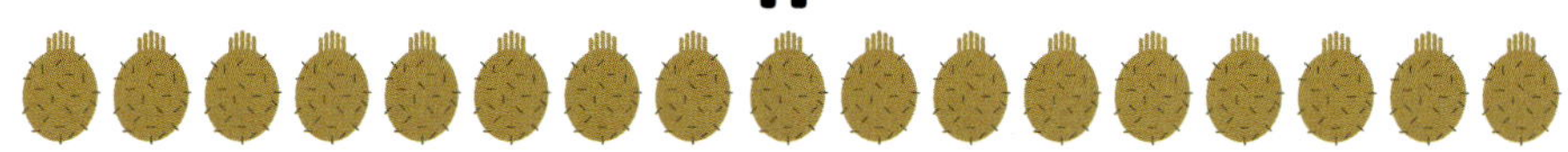

If you know the double of the lower number in an addition sentence, use the doubles + 1 strategy to do mental math to add the numbers.

Double Check

Write the sum for the doubles or near doubles sentence in each box. Try to solve them in your head. Put a check mark (✓) at the end of the row if you used mental math to solve them all. Circle any sentences you might want to practice.

						Check
3 + 4	6 + 7	11 + 11	7 + 8	5 + 4	14 + 14	
8 + 7	10 + 11	4 + 3	10 + 9	12 + 12	12 + 13	
13 + 13	14 + 15	8 + 9	12 + 11	15 + 15	5 + 6	
9 + 8	13 + 14	6 + 5	11 + 10	15 + 16	16 + 16	

Adding Three Numbers

There are a few strategies that make adding three numbers easier. Remember, the commutative property of addition states that you can add in any order.

Make 10

First, look at the numbers in the addition sentence to see if any of the numbers add up to 10.	6 + 8 + 4 = ?
Add the two numbers that make 10.	10 + 8 = ?
Then add the remaining number.	10 + 8 = 18

Doubles Strategy

First, look at the numbers in the addition sentence to see if any of the numbers are doubles.	8 + 3 + 8 = ?
Add the doubles.	16 + 3 = ?
Then add the remaining number.	16 + 3 = 19

Try This!

Circle the numbers in the sentences that make 10. The first one has been done for you.

8 + 16 + 2 = 26

4 + 19 + 6 = 29

3 + 7 + 11 = 21

Try That!

Circle the numbers in the sentences that are doubles. The first one has been done for you.

5 + 17 + 5 = 27

4 + 4 + 8 = 16

2 + 9 + 9 = 20

Encourage your child to use mental math when incorporating strategies like the ones here.

Strategy Session

Write the sum for each of the problems below. Then check the box to show which strategy you used. The first one has been done for you. **Hint:** There will be times when you can use more than one strategy.

6 + (7) + (7) = 20 ☐ Make 10 ■ Doubles	8 + 4 + 2 = ___ ☐ Make 10 ☐ Doubles	2 + 14 + 2 = ___ ☐ Make 10 ☐ Doubles	6 + 4 + 9 = ___ ☐ Make 10 ☐ Doubles
11 + 5 + 11 = ___ ☐ Make 10 ☐ Doubles	3 + 7 + 5 = ___ ☐ Make 10 ☐ Doubles	1 + 8 + 9 = ___ ☐ Make 10 ☐ Doubles	4 + 8 + 8 = ___ ☐ Make 10 ☐ Doubles
7 + 7 + 3 = ___ ☐ Make 10 ☐ Doubles	4 + 4 + 11 = ___ ☐ Make 10 ☐ Doubles	3 + 6 + 3 = ___ ☐ Make 10 ☐ Doubles	6 + 4 + 6 = ___ ☐ Make 10 ☐ Doubles
10 + 9 + 1 = ___ ☐ Make 10 ☐ Doubles	3 + 9 + 9 = ___ ☐ Make 10 ☐ Doubles	12 + 8 + 2 = ___ ☐ Make 10 ☐ Doubles	10 + 7 + 10 = ___ ☐ Make 10 ☐ Doubles

While a given strategy may be more efficient for a particular problem, remind your child that there is often more than one way to solve a problem.

Part-Part-Whole Number Bonds

You can use part-part-whole number bonds to solve addition and subtraction problems.

For addition, add the parts together to equal the whole. You can use different strategies to add.

To find the whole, try the doubles - 1 strategy.

Notice the number 7 is 1 less than 8.	8 + 7
Write the sentence as a doubles - 1.	8 + 8 - 1
Add the doubles together and subtract 1.	16 - 1
The answer is the whole.	15

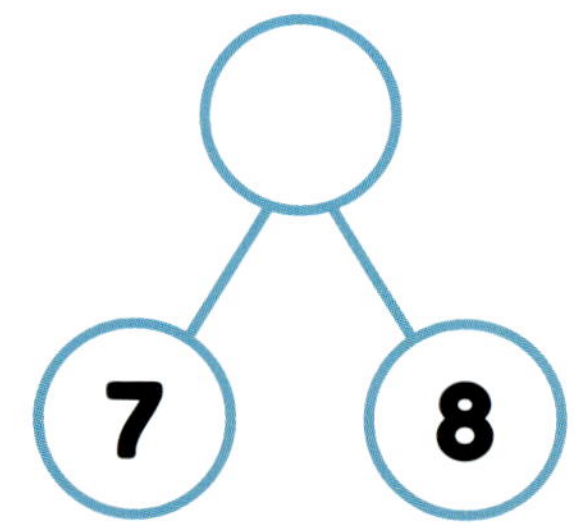

For subtraction, take a part away from the whole to find the missing part. You can use different strategies to subtract.

To find the part, try the counting back strategy.

Count back the part, 2, from the whole, 20.	20 - 2
20 to 19, 19 to 18.	20 - 2 = 18
Whole - part = part!	18

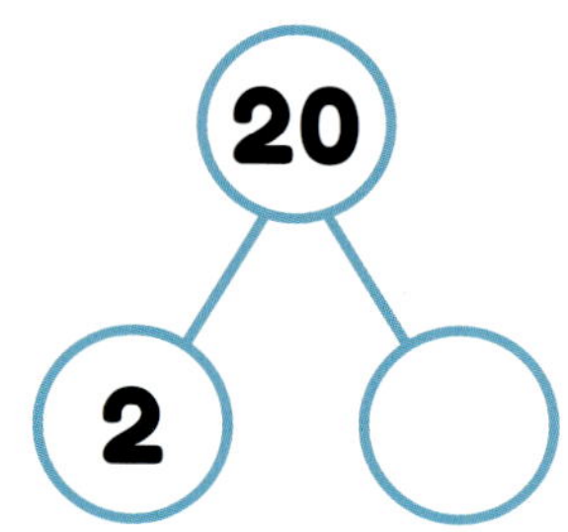

Bonded

Look at the number bonds below. Each one is missing the whole or a part. Below each number bond, write an addition or subtraction sentence and find the unknown.. The first one has been done for you.

Whole: 14; Parts: 7, ?

14 - 7 = 7

Whole: 19; Parts: ?, 8

____

Whole: ?; Parts: 11, 9

____

Whole: 12; Parts: ?, 6

____

Whole: ?; Parts: 15, 5

____

Whole: 18; Parts: 3, ?

____

Whole: ?; Parts: 9, 10

____

Whole: 11; Parts: ?, 5

____

An Odd Riddle

Use the part-part-whole method to find the missing numbers in the Code Key. Then use the Code Key to fill in the letters to solve the riddle below.

Code Key

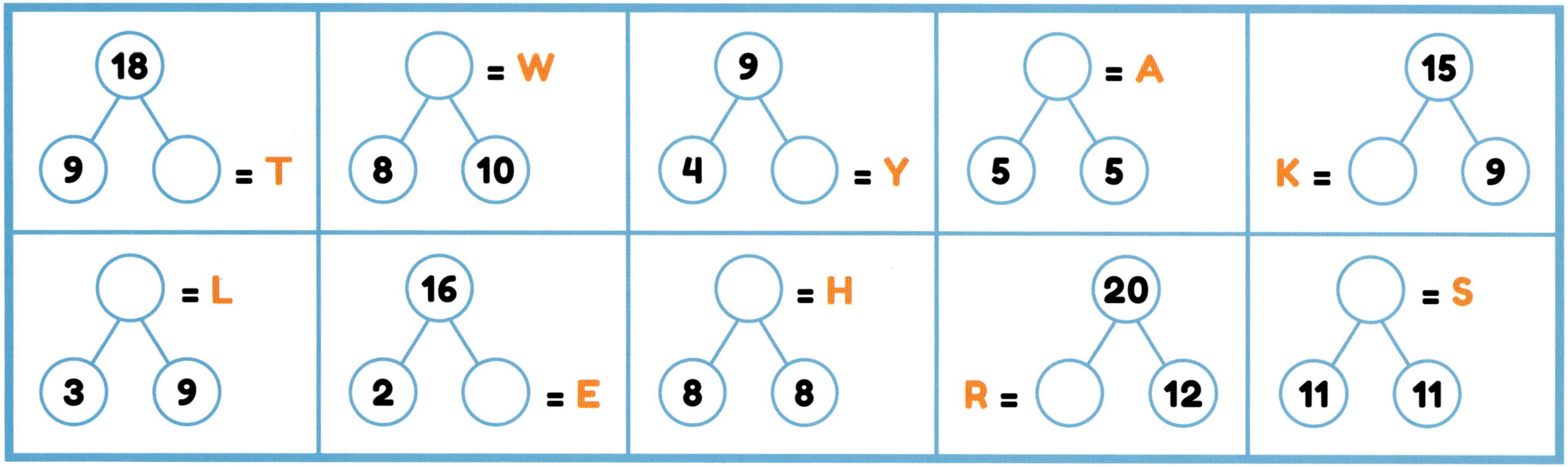

Seven is an odd number. How do you make it even?

___ ___ ___ ___   ___ ___ ___ ___   ___ ___ ___   ___ ___ ___ ___ ___ ___   ___

9 10 6 14   10 18 10 5   9 16 14   12 14 9 9 14 8   22

Pick Up Sticks

Circle the objects to place them into equal groups of 2, 3, 4, or 5. Then write a repeated addition sentence and find the sum. The first one has been done for you. **Hint:** Some groups can be sorted in more than one way.

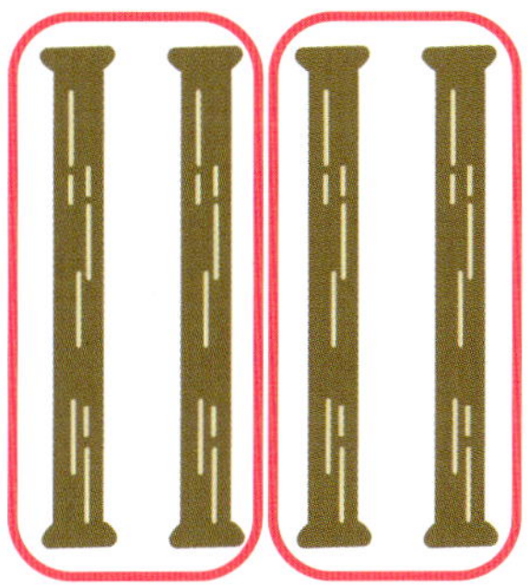

2 + 2 = 4

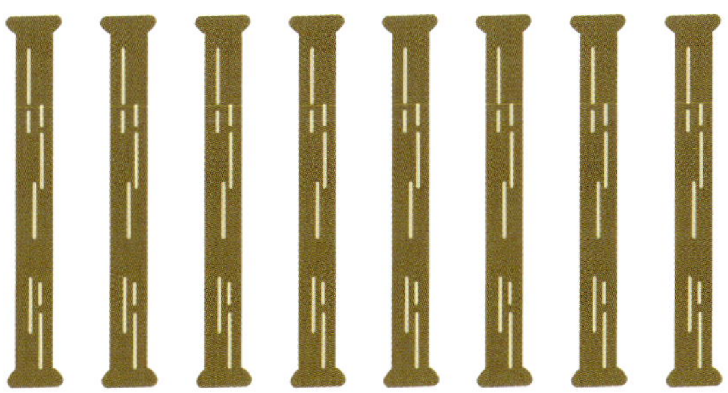

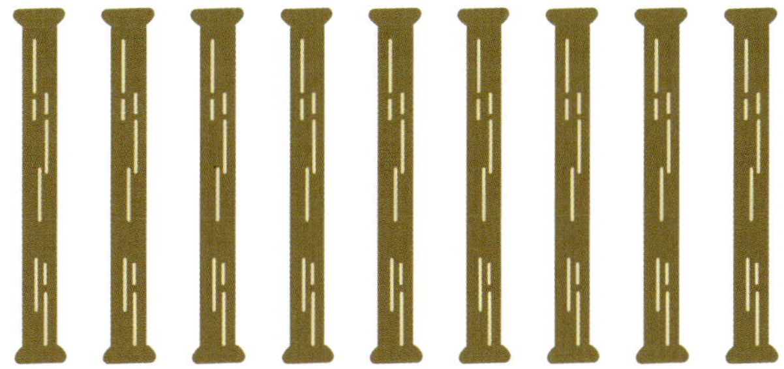

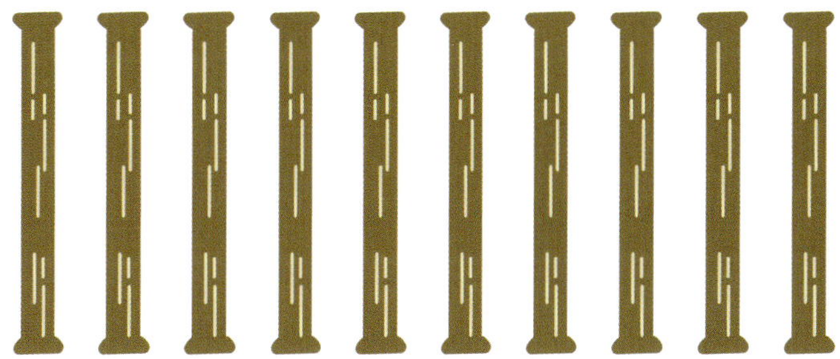

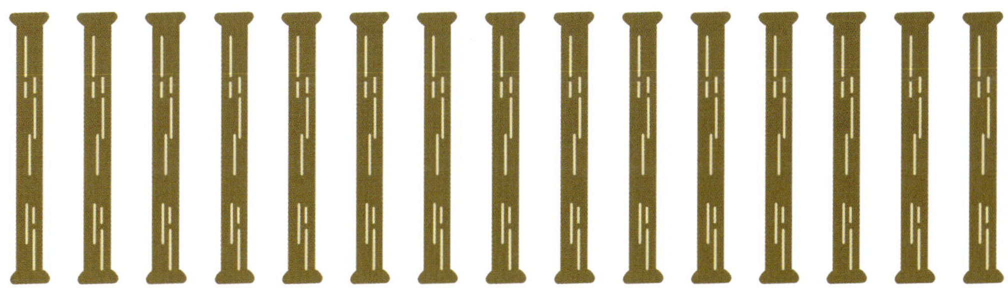

Have a Ball!

Play Ball

Look at the arrays on both pages.
Answer the questions beside each array.
The first one has been done for you.

How many rows? 3

How many in each row? 2

Write the repeated addition sentence and find the sum.

$2 + 2 + 2 = 6$

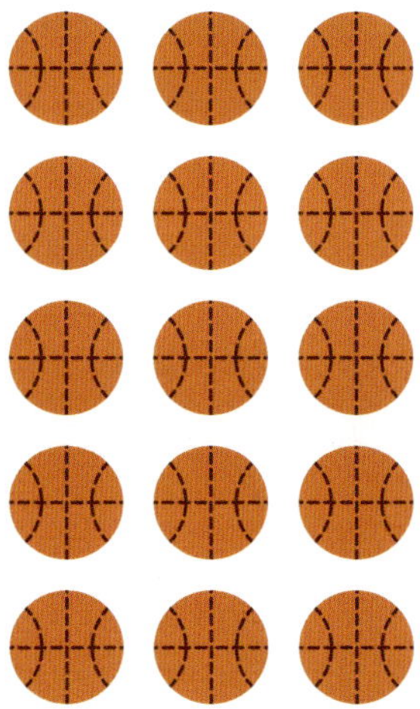

How many columns? ____

How many in each column? ____

Write the repeated addition sentence and find the sum.

____

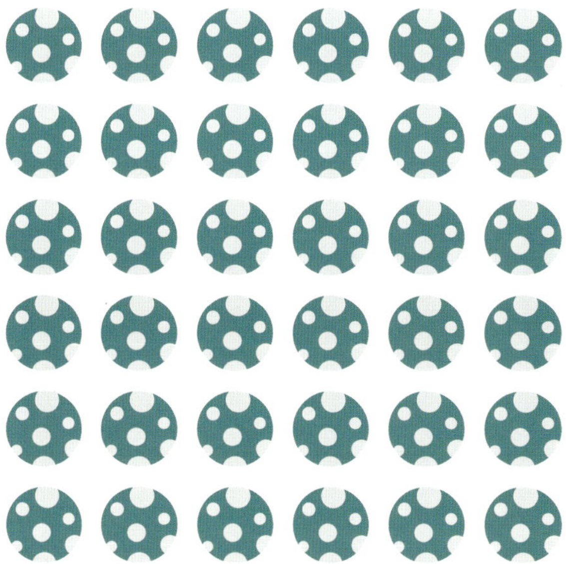

How many rows? ____

How many in each row? ____

Write the repeated addition sentence and find the sum.

____

How many columns? ______

How many in each column? ______

Write the repeated addition sentence and find the sum.

____

How many rows? ______

How many in each row? ______

Write the repeated addition sentence and find the sum.

____

How many rows? ______

How many in each row? ______

Write the repeated addition sentence and find the sum.

____

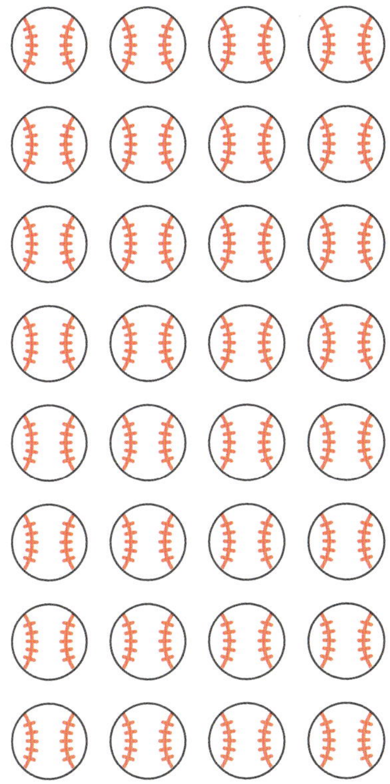

How many columns? ______

How many in each column? ______

Write the repeated addition sentence and find the sum.

____

Little Sister

Your child is ready to read *Little Sister*. This storybook features the near doubles addition strategy. This strategy gives your child another mental math tool to quickly add two numbers.

During reading:

Point out the sets of equations that appear in the book. Occasionally, cover up the second and third lines. Ask your child to identify the doubles + 1 fact.

After reading:

Ask your child how they would feel if they had to share with a sibling or cousin. How would they have handled the situations described in the book?

Play a doubles addition card game!

- Write each doubles equation (without the answer) on an index card.
- Shuffle the cards and lay them face down on a table.
- Your child should flip over one card at a time and call out the sum for that equation.
- When they correctly identify a sum, they can keep the card. If they get the sum wrong, the card goes back on the table, face down.

Celebrate your child's progress by placing a star sticker in the workbook.

Party Time

Color the party balloons! If the balloon has an odd number, color it **blue**. If it has an even number, color it **yellow**.

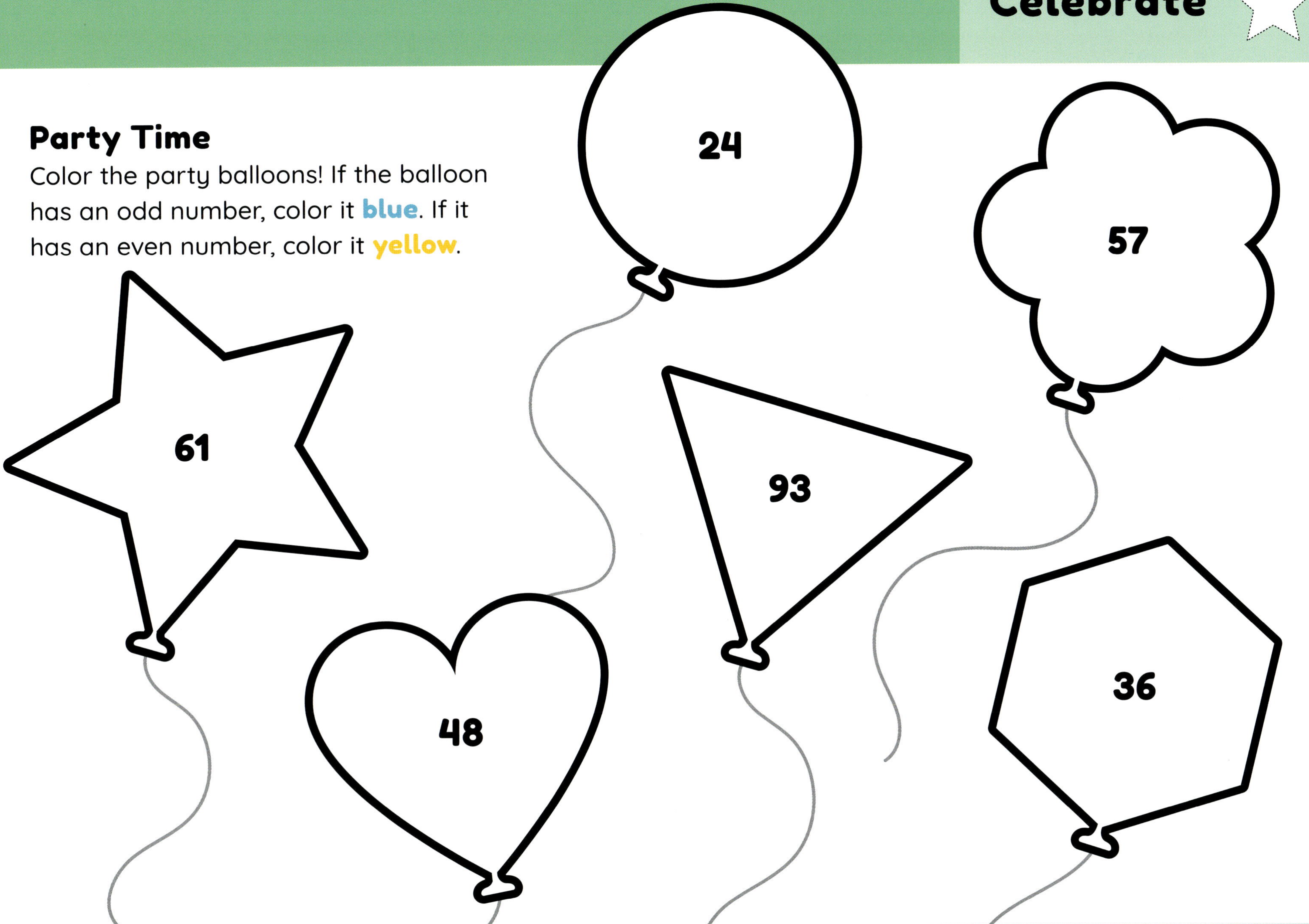

Your child has completed Step 2!
Celebrate by placing a star sticker in the workbook.

Hooked on Math

It's Time to Celebrate!

____

has completed Elementary Math Steps 1 & 2
and is now ready to move on to the
Elementary Math Steps 3 & 4 Workbook!

SKILLS MASTERED

- Identifying even and odd numbers
- Splitting numbers into equal groups
- Writing repeated addition sentences from arrays
- Addition and subtraction strategies practice

Don't forget to share your success and tag us on social media so we can celebrate with you!

Credits

Design Team
Julie Linden Frazier
Lindsey Frazier

Editorial Team
Donna Giachetti
Cynthia Stierle
Austin Feld

Illustrations
Dubit Limited

Educational Advisor
Lesley Freed, Curriculum Specialist

Published by Sandviks HOP, Inc.

Printed in China

Hooked on Math

Elementary Math

18034 (09/22)